RAPID FIRE!

Fast Play World War Two Wargames Rules
For use with Military Miniatures

by Colin Rumford and Richard Marsh

Operation Martlet. The scene at the beginning of move 8 when British air support arrives on table. See page 116 for this brigade-level scenario.

Published by Rapid Fire Publications
1 Willow Park Road
Wilberfoss
York
YO41 5PS
Email: RFpublications@aol.com
Website: http://www.rapid-fire.uk.com

ISBN 0 9550063 0 9

Contents

Rapid Fire!

Rapid Fire! is designed to be a realistic and playable set of rules for WWII wargaming. The complexity of fighting in the mid 20th Century - with its huge range of weaponry, vehicles and 'new' technology - can sometimes result in slow game mechanisms dogged by charts and calculations that try to cover every detail. Rapid Fire! was designed from the beginning to be quick to play, so a game is always challenging, usually enjoyable, sometimes frustrating, but never boring!

The figure scale is designed to give the players the chance to command large units, without losing the 'feel' of combat at lower levels. For this reason each figure represents 15 real troops and each model represents five real guns or vehicles. This may seem strange, but it means that on the average 8' x 5' table Rapid Fire! allows you take charge of up to a regiment (or brigade if British) of three battalions, a battalion of tanks and several artillery batteries, all supported by engineers and aircraft: the sort of force that made a real difference on the WWII battlefield.

Rapid Fire! does not use a strict time and distance scale. Instead, during each move a unit is allowed to carry out a sample of their real-life actions, but in proportion to all other units, (e.g. most mortars can fire faster than most artillery, so mortars may fire twice per move and artillery once). Distances for movement and firing are in proportion to one another and not strictly to scale.

Our aim is simple: to allow large scale battles to be fought on reasonably-sized tables using models big enough to have 'character' when detailed and painted.

Preparing to Play

To play a WWII game using Rapid Fire! the players need two forces of model soldiers, vehicles and guns; a scenario (battlefield situation) to fight; a table-top to play on with suitable terrain to represent the battlefield, and, of course, the rules and playing aids.

The Forces

At the beginning of this book you will find Orders of Battle for units of four of the major nations involved in WWII. As the original Rapid Fire! rule book focused on the early part of WWII we have decided to concentrate on 1944 (the year of D-Day) in the new edition, although the rules – as before – cover the entire conflict. Information for other nations and units can be found in the Rapid Fire! supplements, campaign guides and on our web site.

Which models players use to create these forces is a personal decision. Rapid Fire! games can be played with model soldiers from 15mm – 28mm high (but not at the same time!). The photographs in this book of our own armies feature 20mm figures, vehicles and guns. This scale still offers the greatest choice and the cheapest option if you use easily available plastic figures and models. Don't be discouraged if you feel you can't paint your models like the ones in the photos. Anyone can get better at painting with time, patience and a little bit of inspiration.

The Scenario

Battlefield situations can be created by the players and be as simple as lining up your forces on each side of the table and setting off to eliminate 'the enemy'. However, we recommend historical research and the adaptation of real events to provide scenarios. We would urge all players of Rapid Fire! to read more about the battles they are fighting, to put greater awareness and understanding into their games. This book contains a Basic rules scenario played through step-by-step and a larger game involving use of the Advanced rules.

Table-top and Terrain

Science fiction and wargaming pioneer HG Wells played his games on the floor, but we suggest that you save your knees and use a tabletop. A board, or boards that give a playing surface of 8' x 6' (or 8' x 5') is a good size to aim for, but many games can be played using less space.

Wargames scenery can be as complex or as simple as you like. One of the reasons that we like playing games with models rather than computers is because it's fun to create terrain boards, buildings and the whole battlefield landscape to fight over. However, a green (or desert yellow) cloth will do for starters and the availability of ready-made scenery has never been greater.

The Rules

The Rapid Fire! rules in this book are divided into two sections: Basic and Advanced. The Basic section gives the player every rule they need to play a WWII game involving infantry, tanks and artillery. The Advanced section introduces common, but more complex features of combat, such as engineering and aircraft, and specialised areas such as amphibious and naval warfare. More experienced players will probably dive straight into the Advanced rules and begin using them to add extra interest and challenge to their games. Beginners are advised to familiarise themselves with the Basic set first, then start adding advanced rules gradually. A separate play sheet is included as a quick reference guide.

Playing aids

Players will need homemade plastic fire grid and risk zone templates (see the Rapid Fire! website for help), three D6 die, one D10 dice, a tape measure, a note pad and pen and the various markers explained in the rules.

Acknowledgements

We would like to thank Mark Piper (Australia) for his 'Operation Martlet' scenario and other advice, Chris Leach (Canada), Stephen Bennet (Army Foundation College), Malcolm Taylor, John Durston, Dale Rawson, Howard Greenwood, Dave Johnson, Danny Malone, Tony Chadburn and numerous Rapid Fire! supporters for all their encouragement and help over the past eleven years. We would also like to thank our long-suffering better halves Sue and Jo for all their support and patience.

Colin Rumford and Richard Marsh

British Armoured Brigade - Normandy 1944

- Brigade HQ
  - Armoured Regiment
  - Armoured Regiment
  - Armoured Regiment
  - Motor Battalion
  - Divisional Troops
    - Artillery Regiment

British armoured divisions had one armoured brigade. Independent brigades (including Churchill tank units) were used to supported infantry formations.

Armoured Brigade HQ

6 figures, Humber scout car, Dorchester ACV

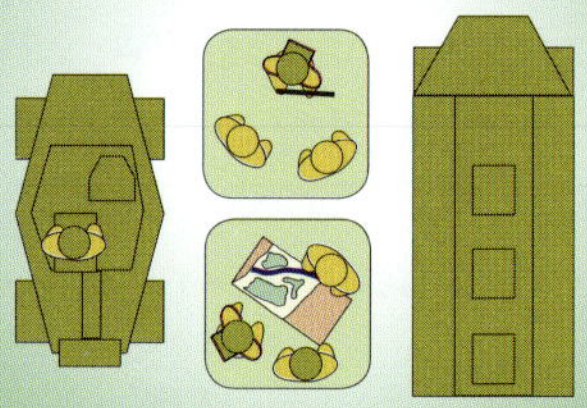

Armoured Regiment

HQ
Humber scout car, Sherman (75mm)
Crusader AA (2 x 20mm), Sherman ARV

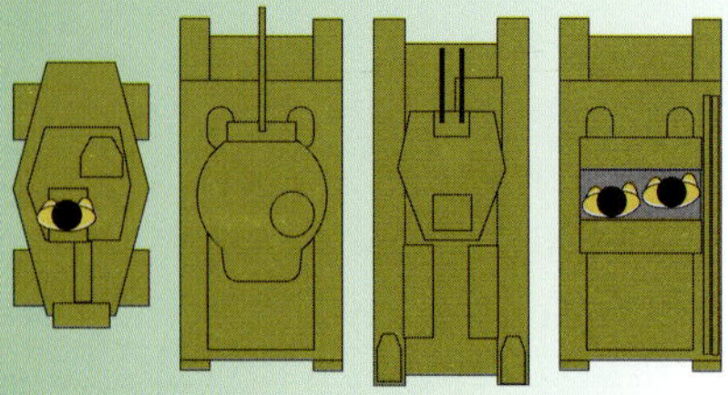

Reconnaissance Platoon
2 x Stuart (37mm)

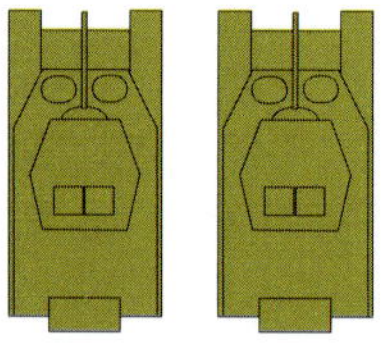

Administrative Platoon
3 x 3 ton ammunition lorry

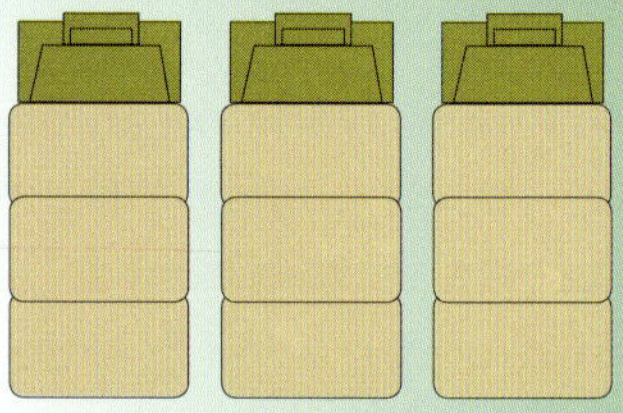

A Squadron
2 x Sherman (75mm), Sherman Firefly (17pdr)

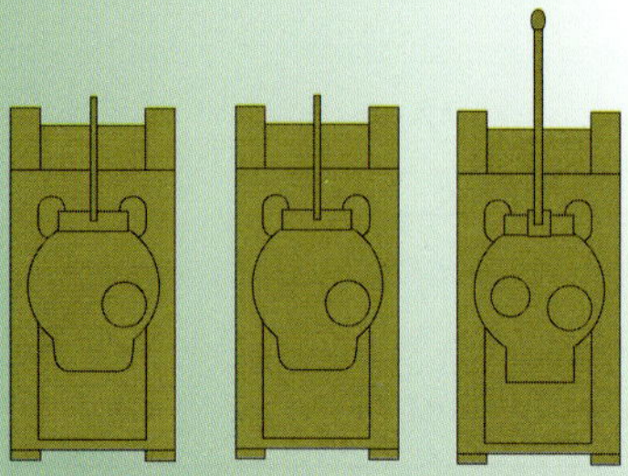

B Squadron
2 x Sherman (75mm), Sherman Firefly (17pdr)

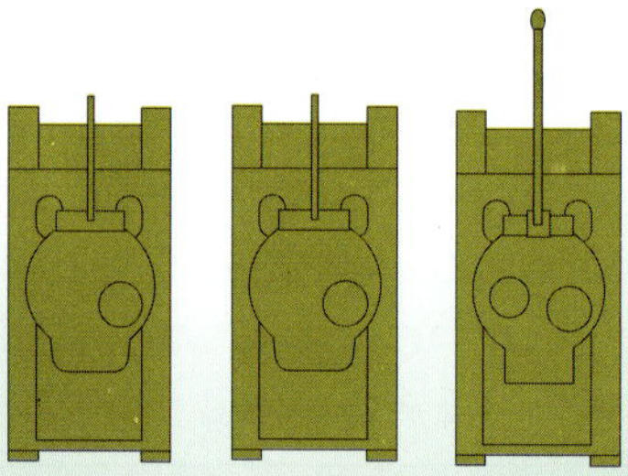

C Squadron
2 x Sherman (75mm), Sherman Firefly (17pdr)

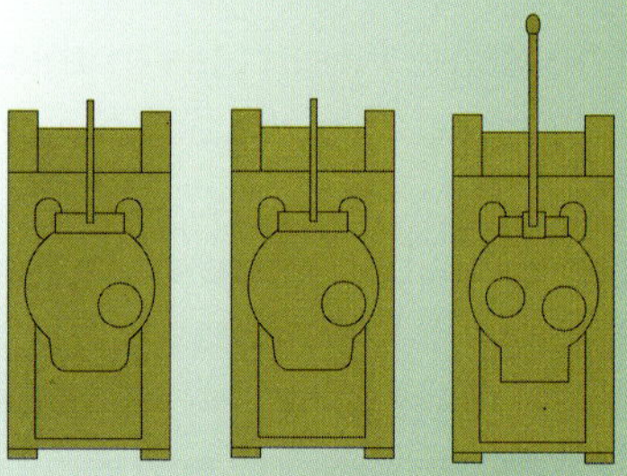

Vehicle and Gun Chart - British Armoured Brigade

Vehicle (Main Weapon)	Speed	Armour	Hull (and Other) Weapons	Turret Weapons
Stuart tank (37mm)	30" / 15"	D	VMG	AT class 4 / HE 6 points table 4 (or VMG)
Sherman tank (75mm)	24" / 12"	C	VMG	AT class 3 / HE 10 points table 4 (or VMG)
Cromwell tank (75mm)	30" / 15"	C	VMG	AT class 3 / HE 10 points table 4 (or VMG)
Sherman Firefly tank (17pdr)*	24" / 12"	C	-	AT class 1 / HE 8 points table 4* (or VMG)
Sherman OP tank	24" / 12"	C	VMG	VMG
Crusader AA tank (2 x 20mm)	30" / 15"	D	-	Multiple autocannon
Sexton SP field gun (25pdr)	24" / 12"	E	AT class 3 / HE 12 points table 4	
Sherman ARV	24" / 12"	C	VMG	
Humber scout car	30" / 15"	E	-	
M3 scout car / M5 half-track (HMG)	30" / 15"	E	HMG	
Bren Carrier	30" / 15"	E	LMG	
Bren Carrier (Vickers MMG)	30" / 15"	E	MMG	
Loyd Carrier	24" / 12"	F	-	
Dorchester ACV	24" / 12"	E	-	
3 ton lorry	24" / 12"	F	-	
Gun				
6pdr anti-tank gun	AT class 3 / HE 6 points table 4**			

* 17pdr AT ranges are:
Short 0 -12",
Medium 12 - 36"
Long 36 - 60".
HE bands are: 6 x 10".
HE is only available from September 1944.

** From July 1944 6pdrs may fire up to two rounds of class 2 ammunition per game (representing APDS ammumition).
HE is only available from September 1944.

Motor Battalion (48 Figures)

HQ
CO + 5 figures, 2" mortar
M3 scout car

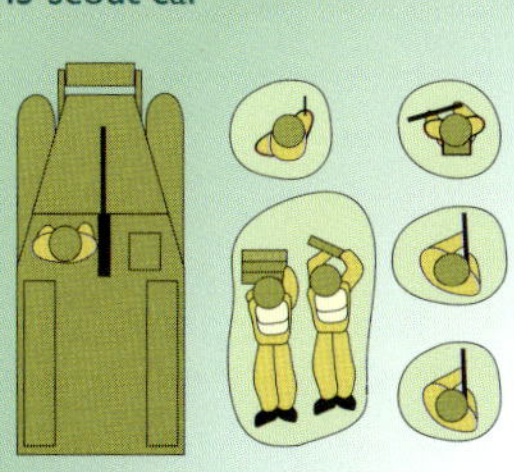

A Company
11 figures, PIAT, M5 half-track, Bren Carrier

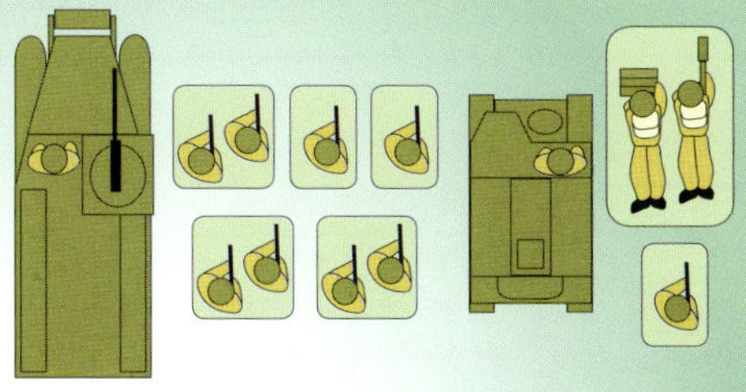

Support Company
3 figures, 6pdr AT gun
Loyd Carrier

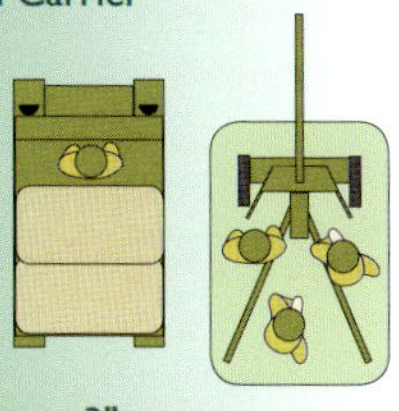

3 figures, 3" mortar
Bren Carrier

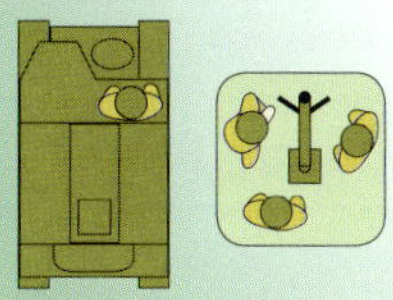

3 figures, Vickers MMG
Bren Carrier

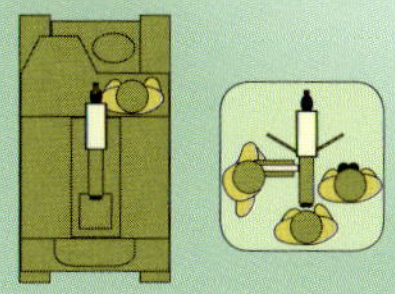

B Company
11 figures, PIAT, M5 half-track, Bren Carrier

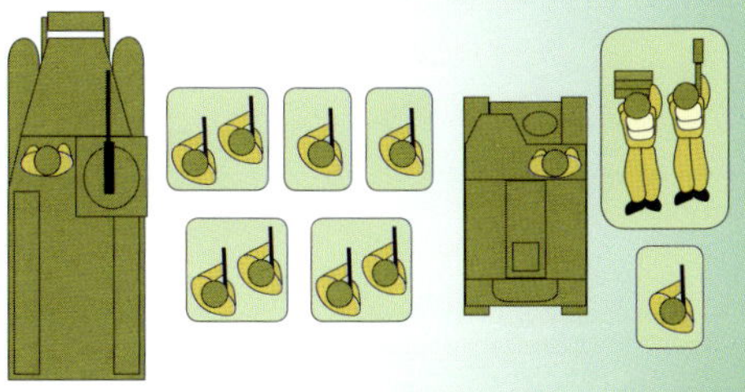

C Company
11 figures, PIAT, M5 half-track, Bren Carrier

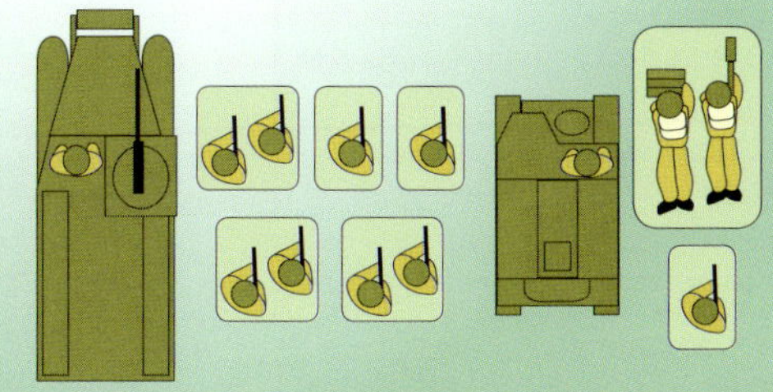

The divisional armoured brigades (and most of the Sherman–equipped independent brigades) had a motor battalion in half-tracks and carriers to provide close infantry support to the tanks and to hold captured ground. A motor company was usually attached to each armoured battalion. A self-propelled (Sexton) 25pdr regiment formed half of every armoured division's artillery and a self-propelled regiment was attached to some of the independent armoured brigades.

Artillery Regiment

HQ
3 figures, M5 half-track, radio truck

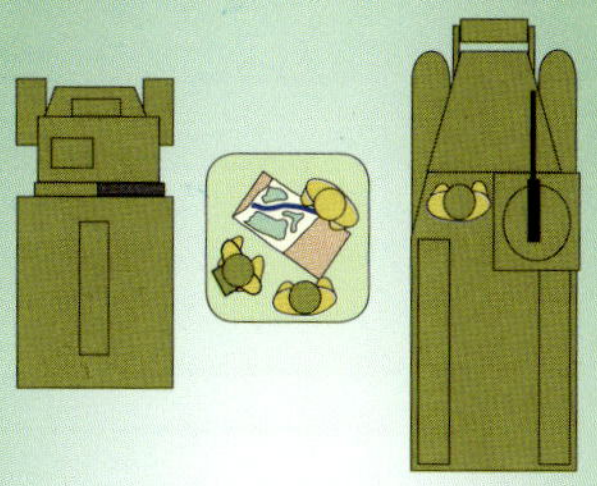

P Battery
2 x Sexton SP (25pdr), Sherman OP tank

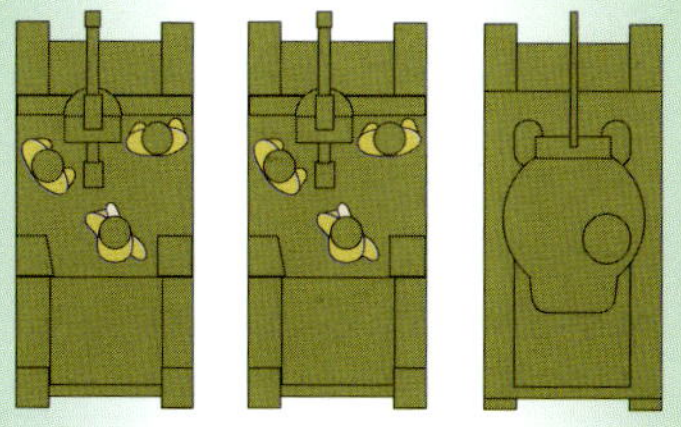

Q Battery
2 x Sexton SP 25pdr field gun, Sherman OP tank

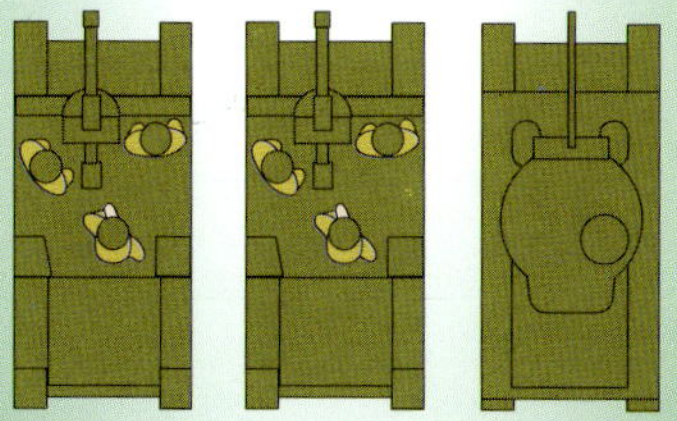

R Battery
2 x Sexton SP 25pdr field gun, Sherman OP tank

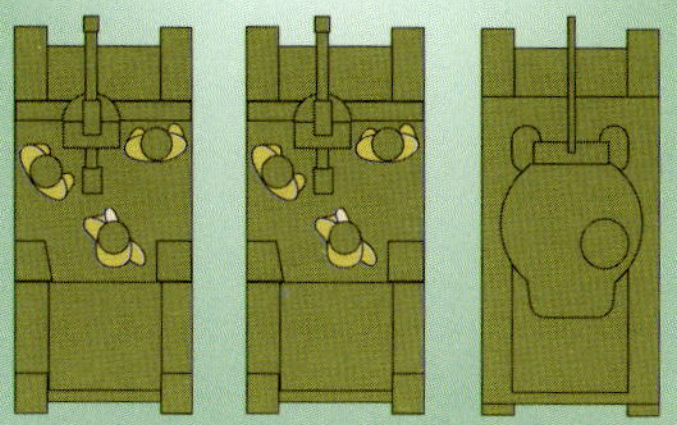

British Armoured Division
Vehicle Serial Numbers and Squadron Markings

	Senior Regt	Second Regt	Junior Regt	Motor Batt	Brigade HQ
Serial Numbers	51	52	53	54	50
HQ / Recce	◇ (red)	◇ (white)	◇ (blue)	◇ (green)	
A Squadron	△ (red)	△ (white)	△ (blue)	△ (green)	
B Squadron	□ (red)	□ (white)	□ (blue)	□ (green)	
C Squadron	○ (red)	○ (white)	○ (blue)	○ (green)	

Squadron Symbols

All British armoured units are normally classed as regular troops.

Options

- 7th Armoured Division armoured regiments were equipped with the Cromwell tank (in place of Sherman 75s). Each Rapid Fire! squadron has two Cromwells and a Sherman Firefly.
- An armoured division had two regiments of field artillery. One was equipped with towed 25 pounders.
- An anti-tank battery may be attached (see the British Infantry Brigade order of battle).

Cromwell (75mm)

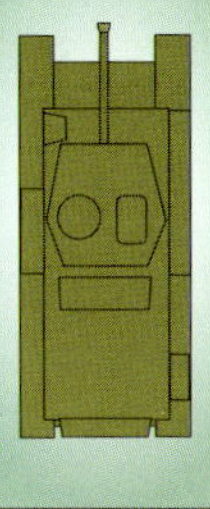

British Infantry Brigade - Normandy 1944

- **Brigade HQ**
  - 1st Battalion
  - 2nd Battalion
  - 3rd Battalion
  - Divisional Troops
    - Artillery Regiment
    - AT Battery
    - MG Company
  - Attached Armour
    - Two Tank Squadrons

Infantry Battalion (48 Figures)

HQ Company
CO + 6 figures, PIAT, 2" mortar

Carrier Platoon
3 figures, PIAT, Bren Carrier

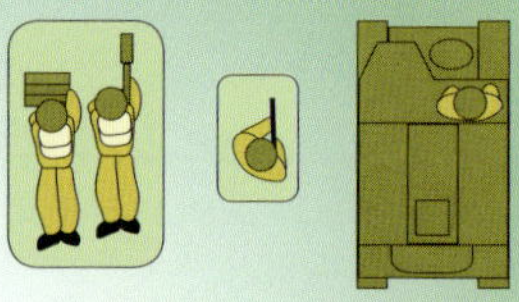

A Company
8 figures

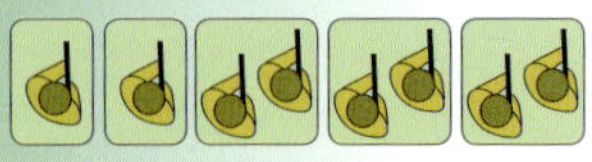

B Company
8 figures

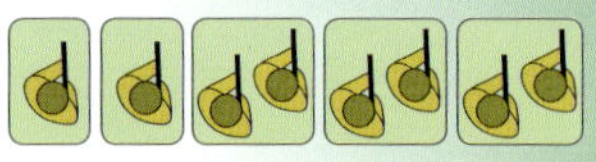

C Company
8 figures

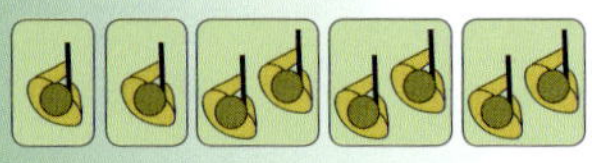

D Company
8 figures

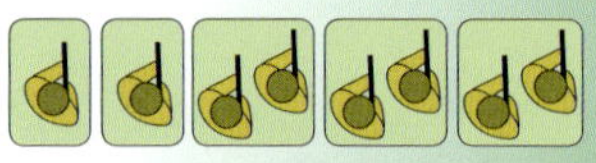

Support Company
3 figures, 6pdr AT gun, Loyd prime mover
3 figures, 3" mortar, Bren Carrier

Motor Pool
15 cwt truck (capacity 8 figures)
2 x 3 ton truck (capacity 16 figures each)

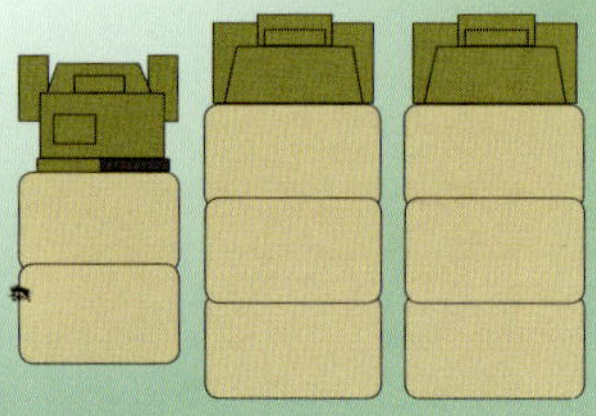

All British infantry units are classed as regular troops.

British Infantry Division Vehicle Serial Numbers

	Senior Brig	2nd Brig	Junior Brig
Brigade HQ	81	87	94
Infantry Battalion	55	60	67
Infantry Battalion	56	61	68
Infantry Battalion	57	62	69
MG Company	64	64	64

Infantry Brigade HQ

6 figures, Humber scout car, 15 cwt radio truck

A British infantry division consisted of three infantry brigades and three field artillery regiments, as well as supporting machine gun, recce, engineer, AT and anti-aircraft units.

Vehicle and Gun Chart - British Infantry Brigade

Vehicle (Main Weapon)	Speed	Armour	Hull (and Other) Weapons	Turret Weapons
Churchill tank (75mm)	18" / 9"	B	VMG	AT class 3 / HE 10 points table 4 (or VMG)
M10 tank destroyer (76mm)	24" / 12"	D	-	AT class 2 / HE 8 points table 4 (or VMG)
Achilles tank destroyer (17pdr)*	24" / 12"	D	-	AT class 1 / HE 8 points table 4* (or VMG)
Sherman OP tank	24" / 12"	C	VMG	VMG
Humber scout car	30" / 15"	E	-	
M5 half-track (HMG) **	30" / 15"	E	HMG	
Bren Carrier	30" / 15"	E	LMG	
Bren Carrier (Vickers MMG)	30" / 15"	E	MMG	
Wasp (flamethrower)	30" / 15"	E	Flamethrower	
15 cwt lorry / radio truck	30" / 15"	F	-	
3 ton lorry, Quad & Loyd prime mover	24" / 12"	F	-	
Gun				
6pdr anti-tank gun	AT class 3 / HE 6 points table 4***			
17pdr anti-tank gun*	AT class 1 / HE 8 points table 4*			
25pdr field gun	AT class 3 / HE 12 points table 4			

* 17pdr AT ranges are:
Short 0 -12",
Medium 12 - 36"
Long 36 - 60".
HE bands are: 6 x 10".
HE is only available from September 1944.

** When towing the 17pdr maximum movement is 24" / 12".

*** From July 1944 6pdrs may fire up to two rounds of class 2 ammunition per game (representing APDS ammumition).
HE is only available from September 1944.

Artillery Regiment

HQ
3 figures, 15 cwt truck, radio truck

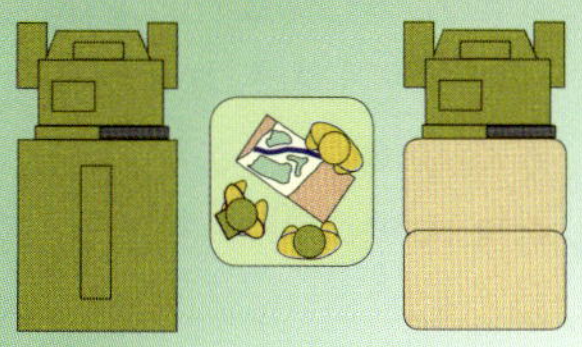

British artillery was plentiful and, by 1944, very flexibly used (see **Firing** and **Specialist Artillery** sections). The MG Company was usually split up between infantry battalions for close support. AT artillery was powerful, but M10s and Achilles were vulnerable to infantry and artillery.

Anti-Tank Battery

M10 (76mm) tank destroyer
Achilles (17pdr) tank destroyer

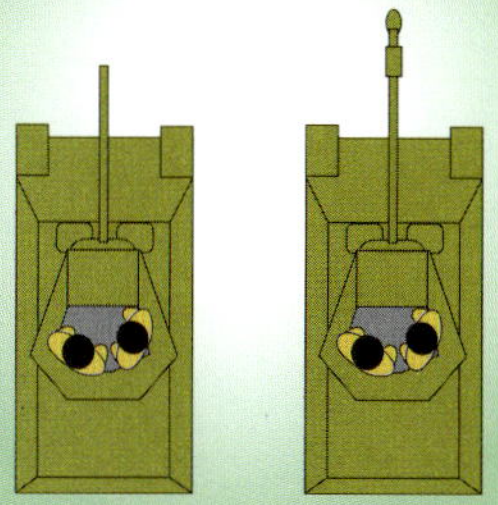

P Battery
10 gunners, 2 x 25pdr field gun, 2 x Quad prime mover, Sherman OP tank

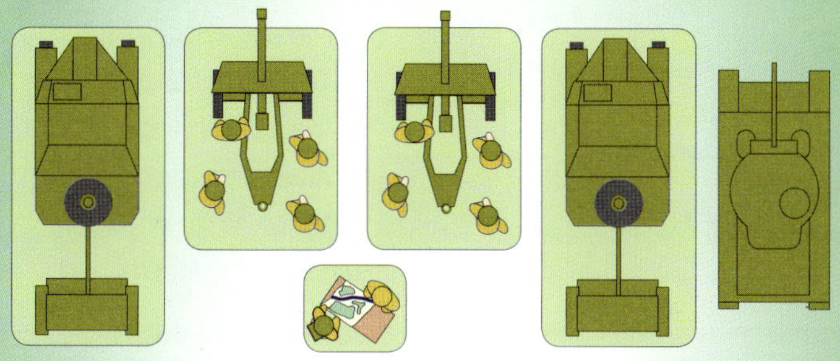

Q Battery
10 gunners, 2 x 25pdr field gun, 2 x Quad prime mover, Sherman OP tank

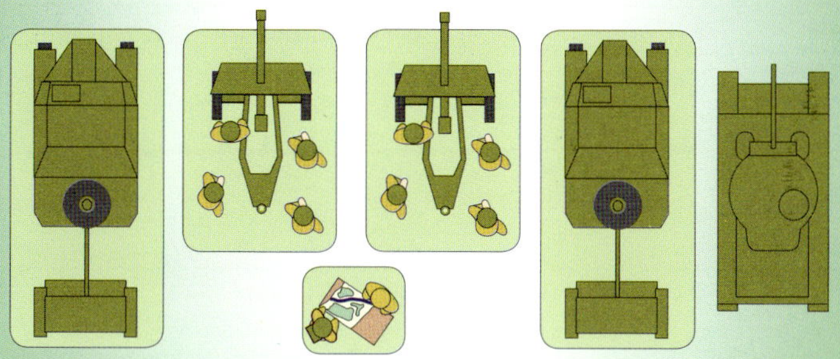

R Battery
10 gunners, 2 x 25pdr field gun, 2 x Quad prime mover, Bren Carrier OP

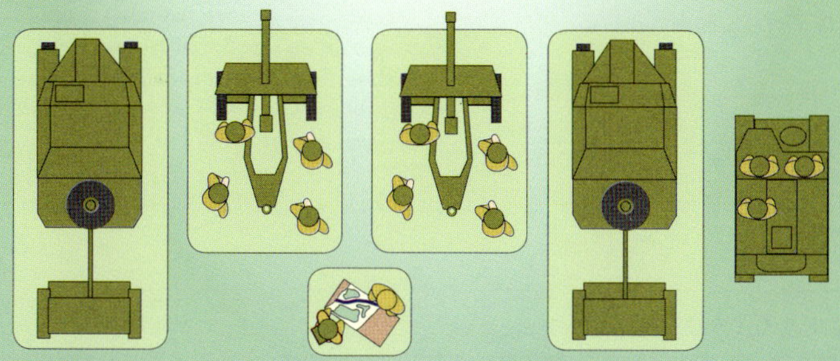

Machine Gun Company (16 Figures)

CO + 3 figures, PIAT, Bren Carrier
3 figures, 4.2" mortar, Loyd Carrier
9 figures, 3 x Vickers MMG, 3 x Bren Carrier

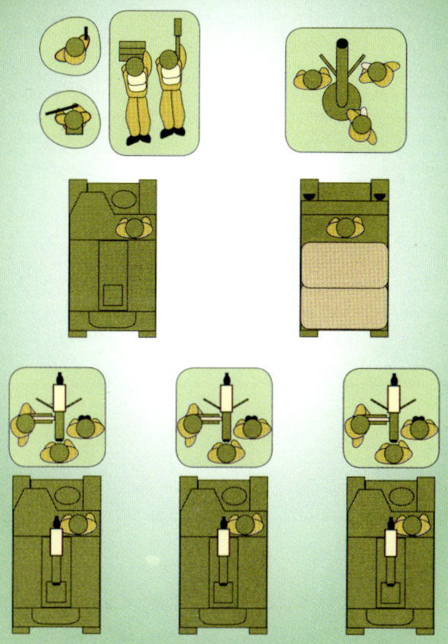

Tank Squadron

3 x Churchill (75mm)

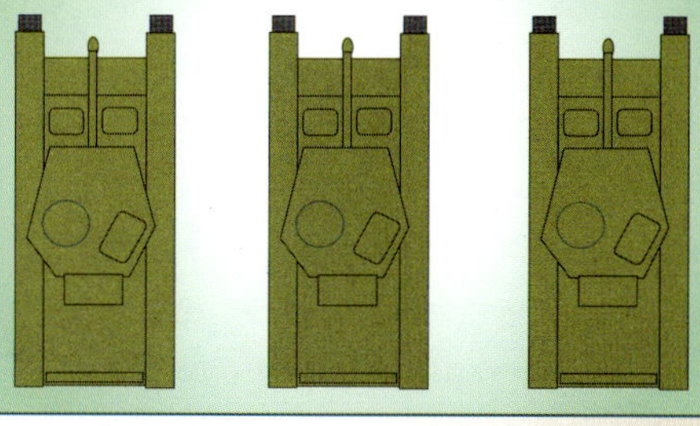

Options

1 Battalions may have a pioneer platoon (two figures with a flamethrower).

2 From July 1944 the MG company may add a Wasp (carrier with flamethrower).

3 The anti-tank company may be equipped with towed 17pdr anti-tank guns.

4 An engineering field company (8 figures, mine clearing equipment and M5 half-tack (HMG)) may be added.

1

2

Wasp

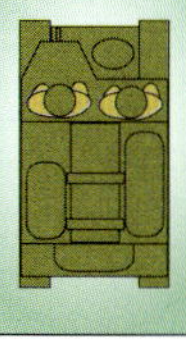

3

8 gunners, 2 x 17pdr AT gun, 2 x M5 half-track prime mover

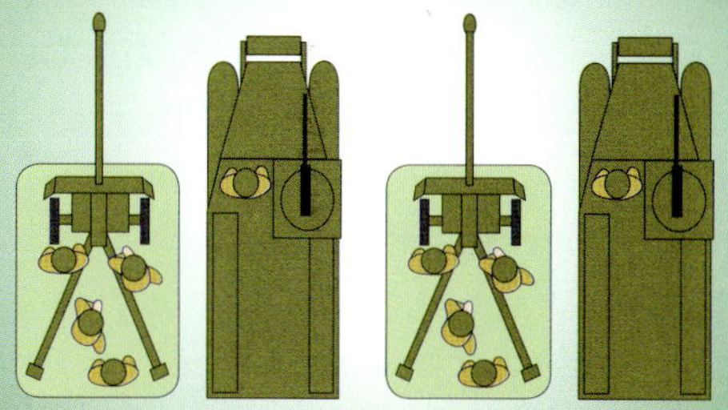

German Infantry Division Kampfgruppe - Normandy 1944

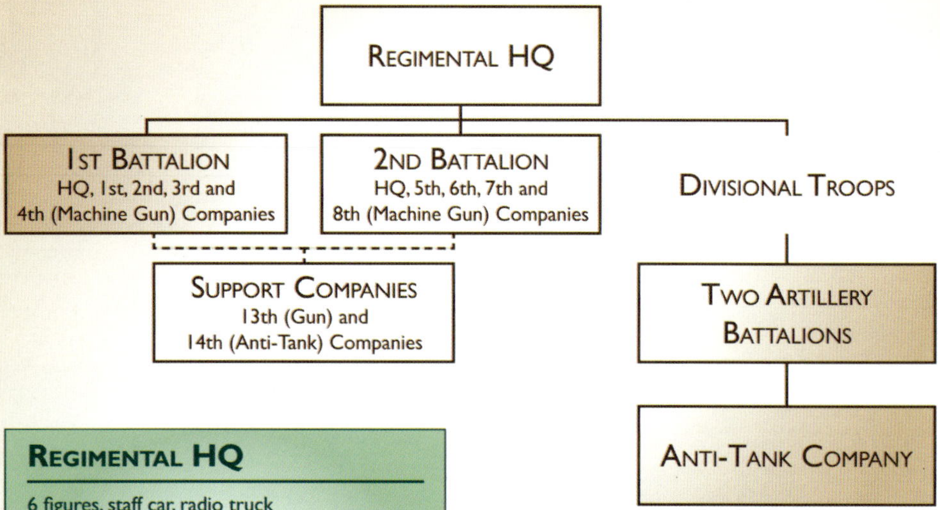

German kampfgruppes were totally flexible grouping of units put together to suit particular tactical situations. These battle groups could be anything from a group of stragglers to a powerful all-arms force, like the one represented here. German troops were trained to switch from unit to unit without losing cohesion.

German infantry are normally classed as regular troops.

Regimental HQ

6 figures, staff car, radio truck

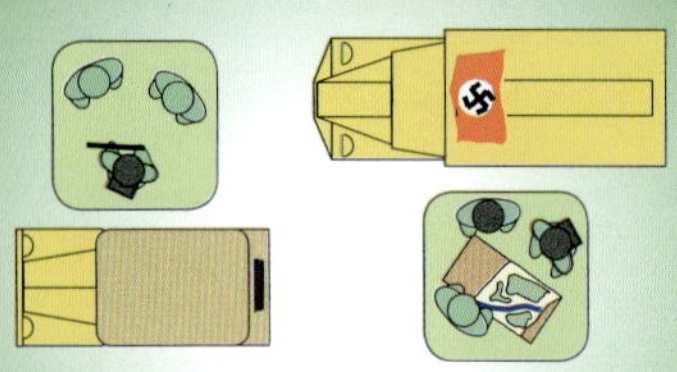

Recce and Engineer Platoons
10 figures (5 are engineers), Panzerfaust

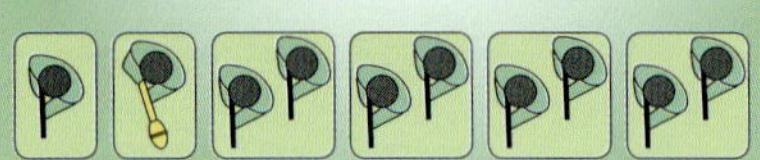

Our kampfgruppe is based around an infantry regiment. It has a small HQ, with a useful reserve in its combined reconnaissance and engineer platoons, two infantry battalions and a support company with close support infantry guns and towed and hand held anti-tank weapons. The two battalions (reduced from earlier in the war) were 'foot-sloggers' with no motor transport, but we attach vehicle-towed weapons from the support companies for morale purposes. By this stage of the war the lack of transport wasn't such a great disadvantage as the infantry units were spending most of their time in defence. Hand–held anti-tank weapons are generously distributed to counter the increasing number of Allied tanks.

Infantry Battalion (39 Figures)

HQ
CO + 5 figures

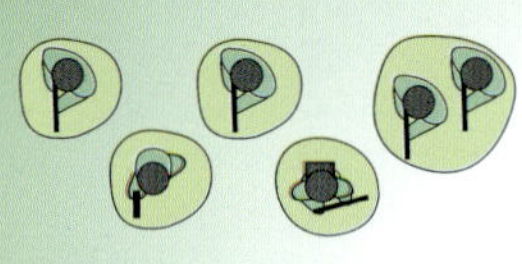

1st Company
8 figures, Panzerfaust

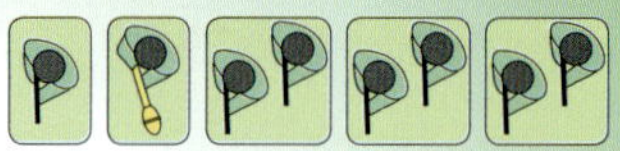

2nd Company
8 figures, Panzerfaust

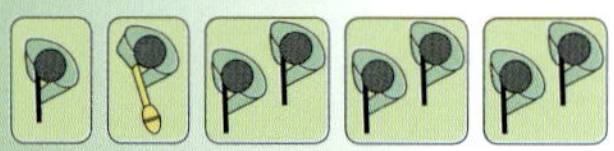

3rd Company
8 figures, Panzerfaust

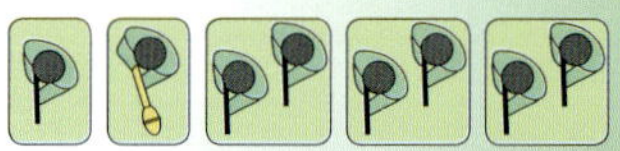

4th (Heavy) Company
9 figures, 2 x MMG, 81mm mortar

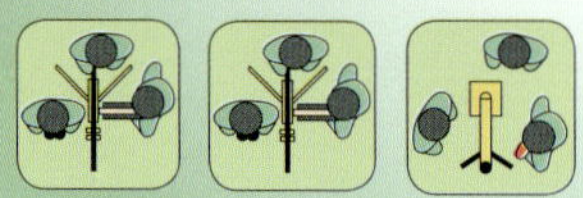

Note: elements from the support companies are attached to each battalion for morale purposes e.g. 1st Battalion may have two Panzerschrecke teams and the 150mm infantry gun attached, which gives a starting morale strengh of 46 (39 + 4 + 3).

Support Companies (17 Figures)

13th (Gun) Company
3 figures, 150mm infantry gun
Medium truck prime mover

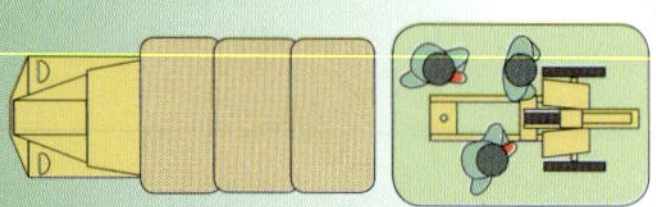

3 figures, 75mm infantry gun
Light truck prime mover

14th (Anti-Tank) Company
8 figures, 4 x Panzerschrecke

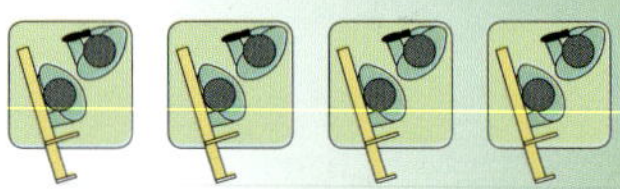

3 figures, 75mm (PaK 40) anti-tank gun
Medium truck prime mover

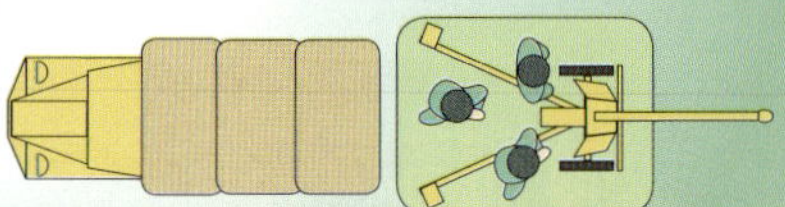

Two Battalions from the Divisional Artillery Regiment

HQ
3 figures, radio truck

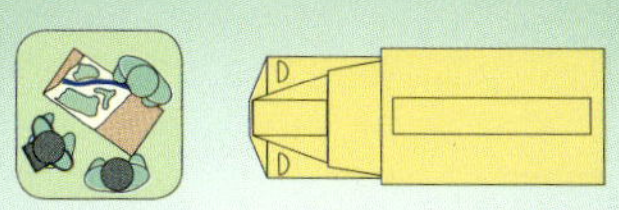

1st Battalion
10 gunners, 2 x 105mm howitzer, 2 x horse drawn limber, 2 figure OP team, light car

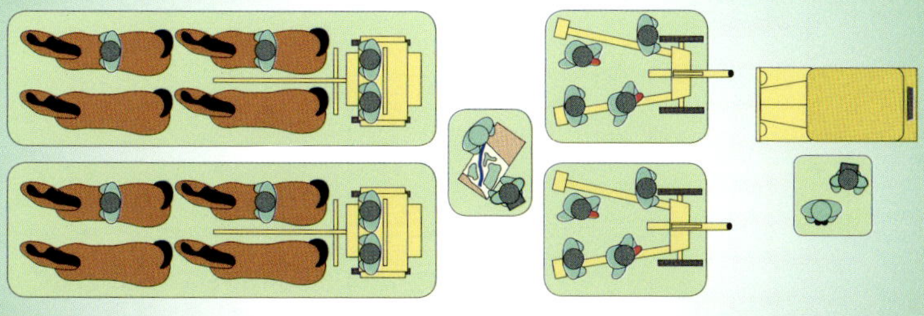

2nd Battalion
10 gunners, 2 x 105mm howitzer, 2 x horse drawn limber, 2 figure OP team, light car

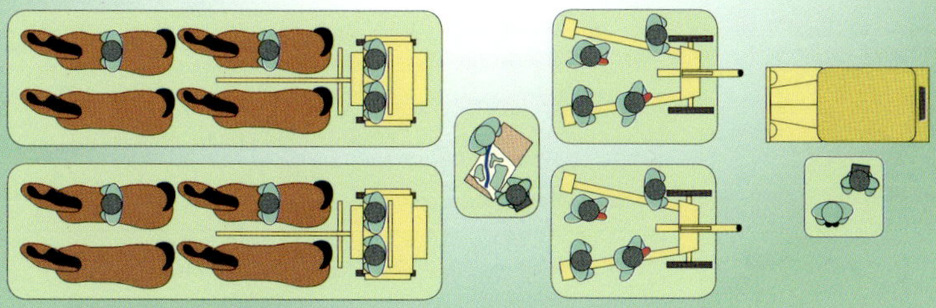

Anti-Tank Company

2 x Marder III (75mm)

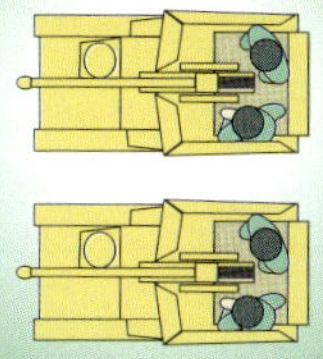

German infantry on the march. Most German troops travelled this way in 1944!

Vehicle and Gun Chart - German Infantry Kampfgruppe

Vehicle (Main Weapon)	Speed	Armour	Hull (and Other) Weapons
StuG III assault gun (75mm)	24" / 12"	C	AT class 2 / HE 10 points table 4
Marder III tank destroyer (75mm)	24" / 12"	E	AT class 2 / HE 10 points table 4
Staff car / light truck	30" / 15"	F	-
Medium truck / prime mover	24" / 12"	F	-
Horse drawn limber	15" / 9"	F	-
Gun			
75mm infantry gun	AT class 5 / HE 10 points table 4		
75mm (PaK 40) anti-tank gun	AT class 2 / HE 10 points table 4		
105mm howitzer	AT class 3 / HE 12 points table 4		
150mm infantry gun	AT class 3 / HE 12 points table 5		

Options

1 One battalion may have an additional 120mm mortar.

2 The anti-tank company may be equipped with StuG IIIs or towed PaK 40s (or in some units towed 37mm AA guns due to equipment shortages).

1

3 figures
120mm mortar
Light truck

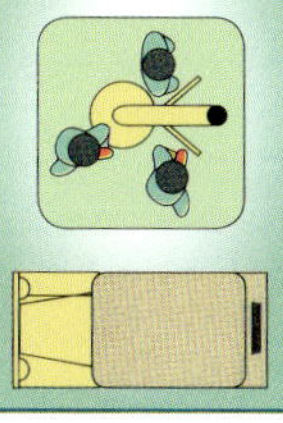

2

2 x StuG III (75mm)

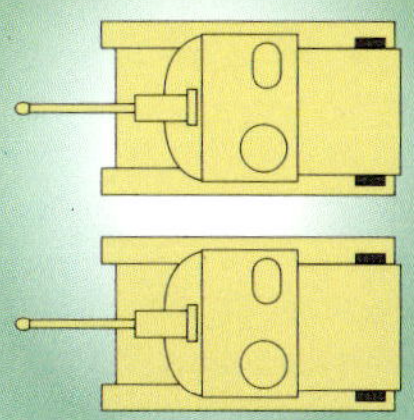

8 gunners, 2 x 75mm (PaK 40) AT gun
2 x medium truck

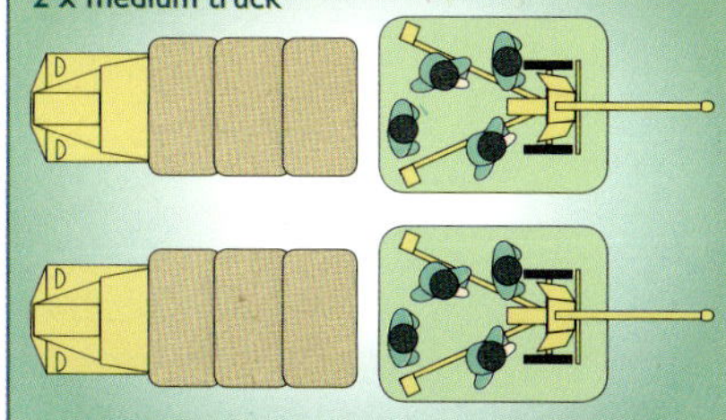

German Panzer Division Kampfgruppe - Normandy 1944

- **Regimental HQ**
  - **Panzer Grenadier Regiment**
    - **1st Battalion (Armoured)** — HQ, 1st, 2nd, 3rd and 4th (Heavy) Companies
    - **2nd Battalion (Motorised)** — HQ, 5th, 6th, 7th and 8th (Heavy) Companies
    - **Support Companies** — 13th (Gun) Company, 16th (Pioneer) Company
  - **Divisional Troops**
    - **Panzer Battalion**
    - **Anti-Tank Company**
    - **Artillery Battalion**

Regimental HQ

3 figures, staff car
3 figures, Sdkfz 251/3 half-track

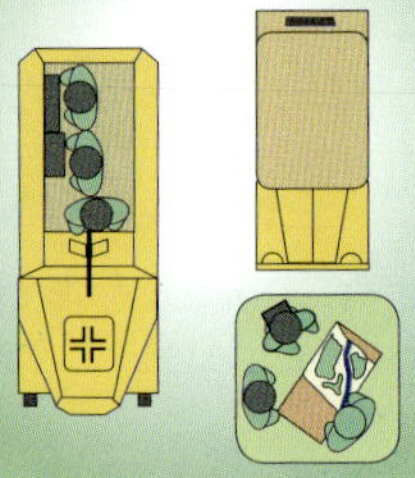

The panzer grenadier regiment was a well-armed force designed to support and fight with tanks. Only Panzer Lehr Division had all its panzer grenadier battalions mounted in armoured half-tracks.

Armoured Panzer Grenadier Battalion

(44 figures*)

HQ
CO + 4 figures, Panzerschrecke
Sdkfz 251/10 half-track (37mm)

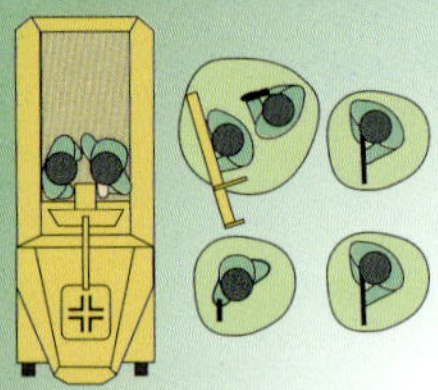

1st Company
8 figures, Panzerfaust
Sdkfz 251/1 half-track

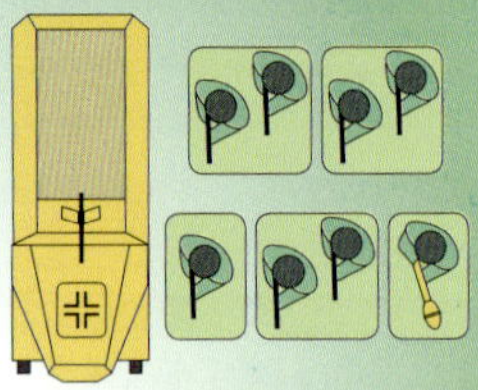

2nd Company
8 figures, Panzerfaust
Sdkfz 251/1 half-track

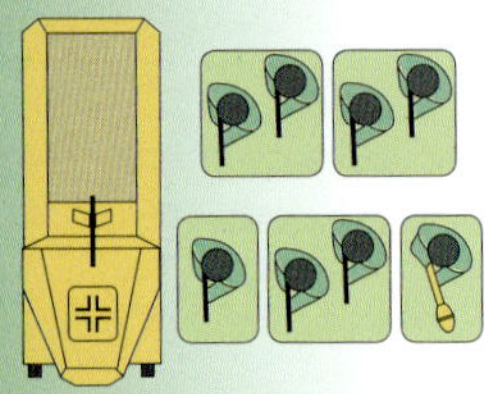

3rd Company
8 figures, Panzerfaust
Sdkfz 251/1 half-track

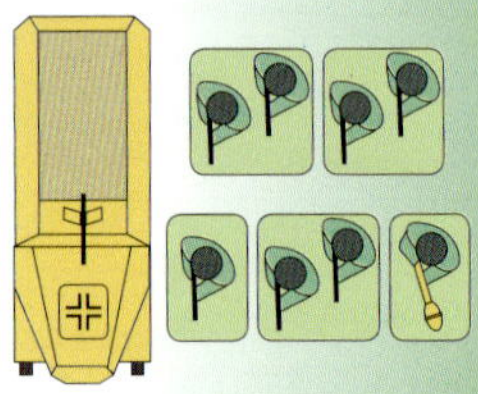

4th (Heavy) Company
6 figures, 2 x MMG, Sdkfz 251/1half-track
Sdkfz 251/9 half-track (75mm)
3 figures, Sdkfz 251/2 half-track (81mm mortar)

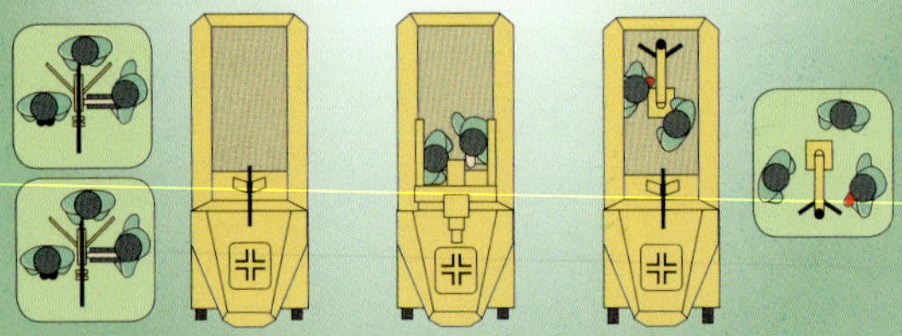

* The Sdkfz 251/10 (37mm) and Sdkfz 251/9 (75mm) half-tracks are classed as AFVs and have a nominal crew of three when counting casualties for morale purposes.

Note: elements from the support companies may be attached to each battalion for morale purposes e.g. the 1st Battalion may have the SP 150mm infantry gun attached which gives a starting morale strengh of 47 (44 + 3).

Motorised Panzer Grenadier Battalion

(41 Figures)

HQ
CO + 4 figures, Panzerschrecke
Personnel carrier

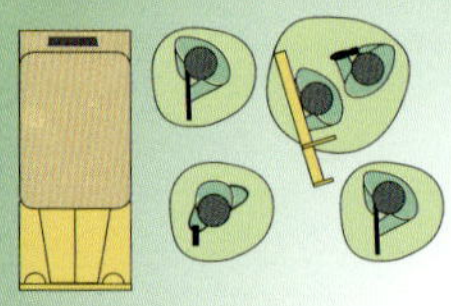

5th Company
8 figures, Panzerfaust
Medium truck

6th Company
8 figures, Panzerfaust
Medium truck

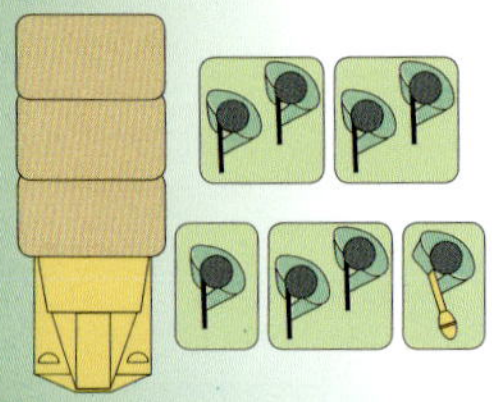

7th Company
8 figures, Panzerfaust
Medium truck

8th (Heavy) Company
3 figures, 120mm mortar, Sdkfz 10 half-track
6 figures, 2 x MMG, Panzerfaust, medium truck
3 figures, 20mm towed AA gun, Sdkfz 10 half-track

Support Companies (14 Figures)

13th (Gun) Company
Grille SP (150mm)

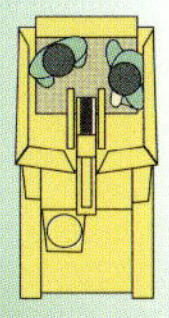

16th (Pioneer) Company
8 figures, Panzerfaust, flamethrower
Sdkfz 251/7 half-track
Sdkfz 251/16 (flamethrower) half-track

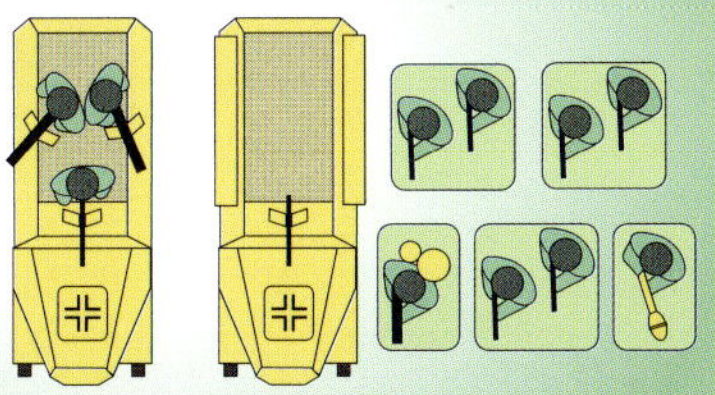

German panzer grenadiers are normally classed as regular troops, but certain units (such as the Panzer Lehr Division and some SS formations) may be elite.

The photograph shows the 15th (Reconnaissance) Company of an SS panzer grenadier regiment. Its Schwimmwagen amphibious cars enabled the unit to cross rivers unaided, useful when probing forward.

Vehicle and Gun Chart - German Panzer Division Kampfgruppe

Vehicle (Main Weapon)	Speed	Armour	Hull (and Other) Weapons	Turret Weapons
Panther tank (75mm)	30" / 15"	B	VMG	AT class 1 / HE 8 points table 4 (or VMG)
Panzer IV tank (75mm)	24" / 12"	C	VMG	AT class 2 / HE 10 points table 4 (or VMG)
Jagdpanzer IV tank destroyer (75mm)	24" / 12"	B	AT class 2 / HE 10 points table 4 (or VMG)	
Bergepanther ARV	30" / 15"	B	VMG	
Sdkfz 251/1/3/7 and 250/5 half-tracks	30" / 15"	E	LMG	
Sdkfz 251/2 half-track (mortar)*	30" / 15"	E	LMG or 81mm mortar	
Sdkfz 251/9 half-track (75mm)	30" / 15"	E	AT class 3 / HE 10 points table 4	
Sdkfz 251/10 half-track (37mm)**	30" / 15"	E	AT class 4 / HE 6 points table 4	
Sdkfz 251/16 (flamethrower) half-track	30" / 15"	E	Flamethrower	
Grille SP infantry gun (150mm)	24" / 12"	E	AT class 3 / HE 12 points table 5	
Wespe SP howitzer (105mm)	24" / 12"	E	AT class 3 / HE 12 points table 4	
Staff car / Schwimmwagen***	30" / 18"	F	-	
Medium truck / prime mover#	24" / 12"	F	-	
Sdkfz 10 half-track / Opel Maultier#	30" / 15"	F	-	
Gun				
75mm (PaK 40) anti-tank gun	AT class 2 / HE 10 points table 4			
20mm AA gun	20mm autocannon			

* Mortar may fire from the vehicle or be dismounted.

** Can fire hollow charge rocket AT class 2 / 12" range: Short 0 - 4" Medium 4 - 8" Long 8 - 12"). HE 8 points table 4.

*** A Schwimmwagen can cross water obstacles at 4" per turn.

\# Maximum prime mover speed when towing a gun is 24" / 12".

Options

- SS panzer regiments may have two additional support companies (14th - anti-aircraft and 15th - reconnaissance).
- SS panzer grenadier regiments normally have three battalions.
- The 4th and 8th (Heavy) Companies may be equipped with a towed 75mm (PaK 40) anti-tank gun.
- The anti-tank company may be equipped with towed 75mm (PaK 40) anti-tank guns (3 crew).

Additional Support Companies (14 Figures)

14th (AA) Company
6 figures, 2 x 20mm towed AA gun
2 x Sdkfz 10 half-track

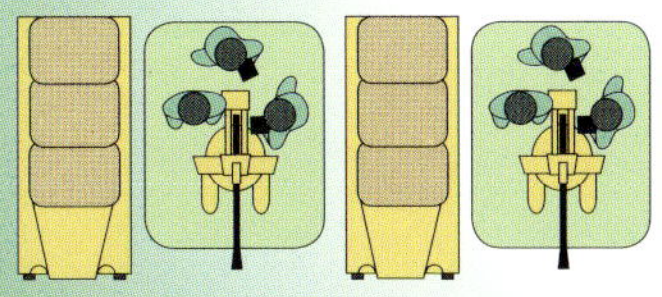

15th (Reconnaissance) Company
8 figures, Panzerfaust
2 x Schwimmwagen

Panzer Division Kampfgruppe Attached Units

Panzer Battalion

HQ
Panther (75mm)
Flakpanzer 38t (20mm), Bergepanther

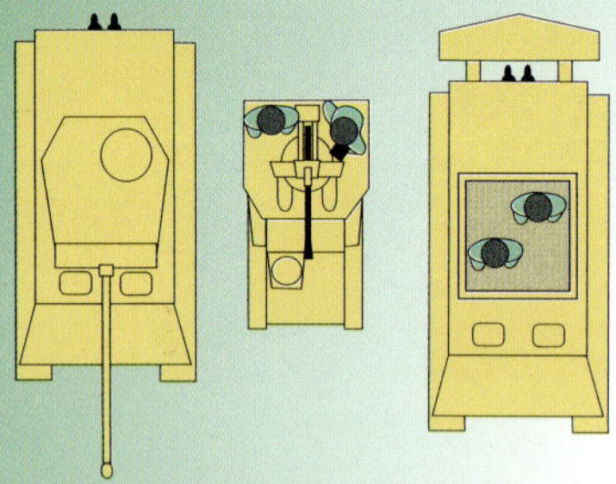

1st Company
3 x Panther (75mm)

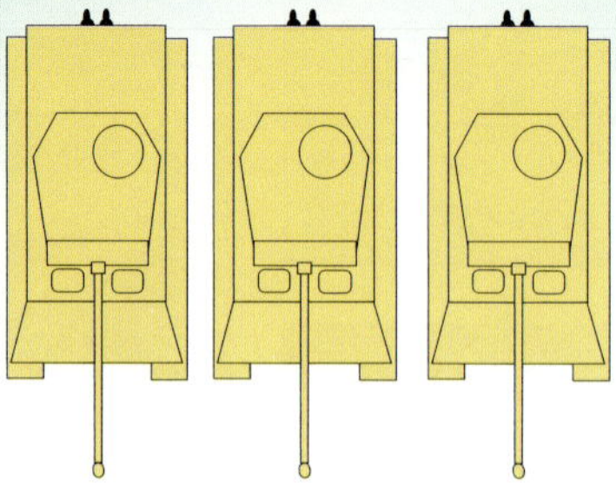

2nd Company
3 x Panther (75mm)

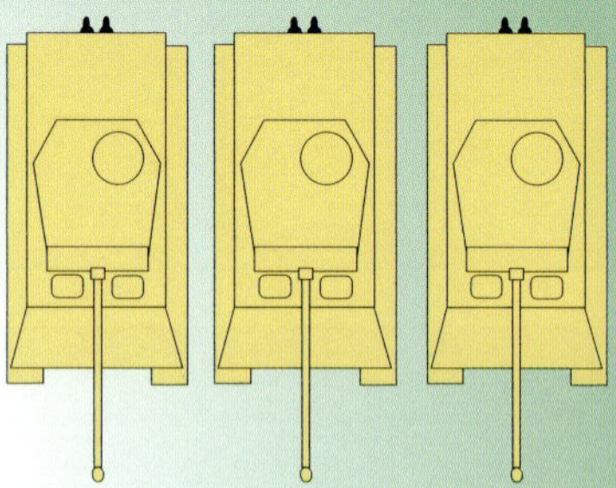

Supply Company
2 x Opel Maultier half-track lorry

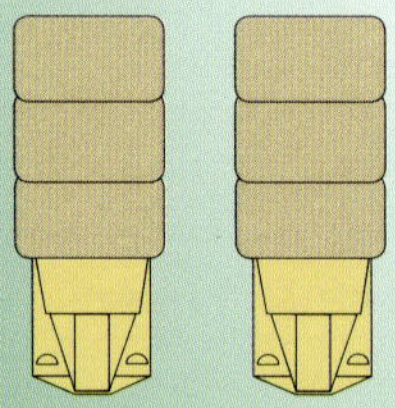

3rd Company
3 x Panther (75mm)

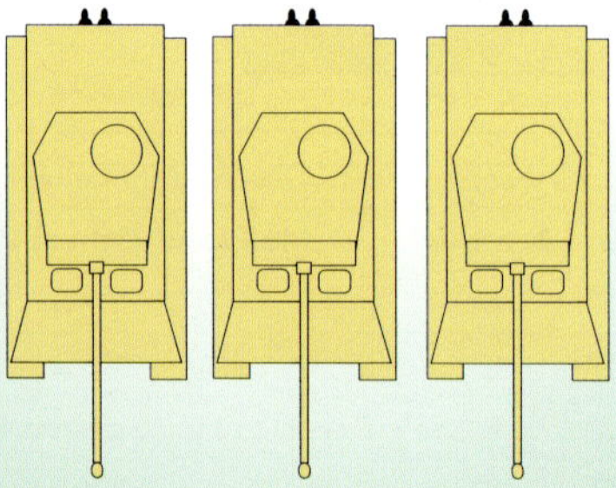

4th Company
3 x Panther (75mm)

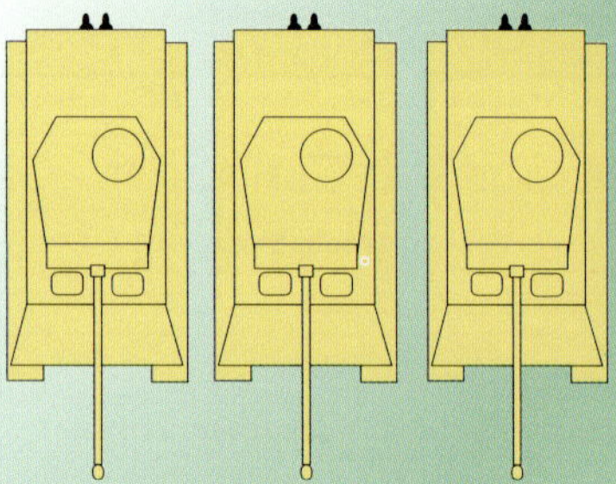

Anti-Tank Company

2 x Jagdpanzer IV (75mm)

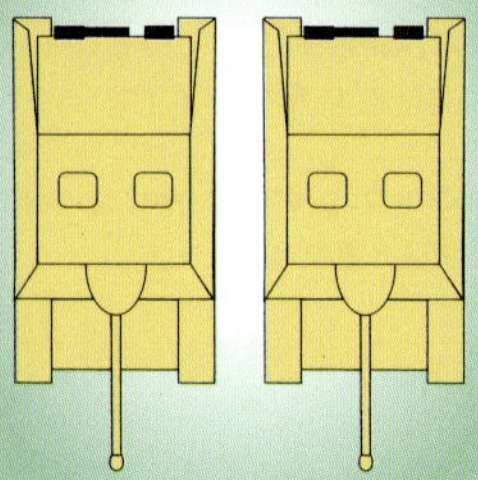

Artillery Battalion*

Hummel SP gun (150mm)
2 x Wespe SP gun (105mm)
Sdkfz 250/5 OP half-track

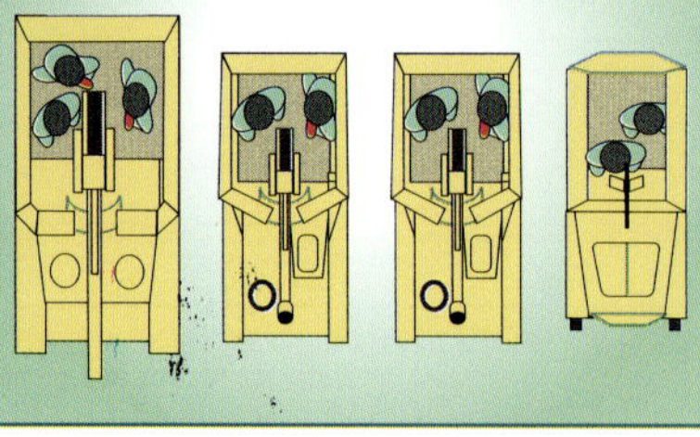

The panzer battalion was a very powerful unit, especially if equipped with Panthers. However, at this stage of the war tanks were difficult to replace. To reflect this, the kampfgruppe may reduce its tank element to one or two companies. If towed panzer artillery battalions are fielded, in 1944 they normally consist of two 105mm (or 150mm) howitzers towed by three (or eight) ton half-tracks.

Options

1 Panzer regiments consist of two battalions. The 1st (companies 1 to 4) is equipped with Panthers. The 2nd (companies 5 to 8) has Panzer IVs.

2 The anti-tank company may be equipped with towed 75mm (PaK 40) anti-tank guns.

* The divisional artillery regiment has three battalions. Normally only one of these is self-propelled, the others are equipped with towed guns.

1 Panzer IV (75mm)

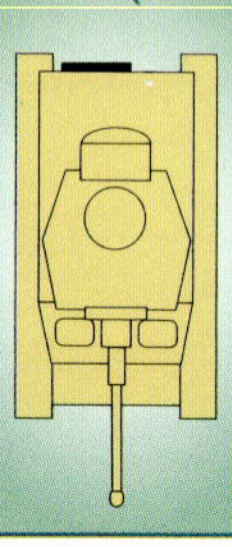

2 8 gunners, 2 x 75mm (PaK 40) anti-tank gun
2 x Sdkfz 11 half-track prime mover

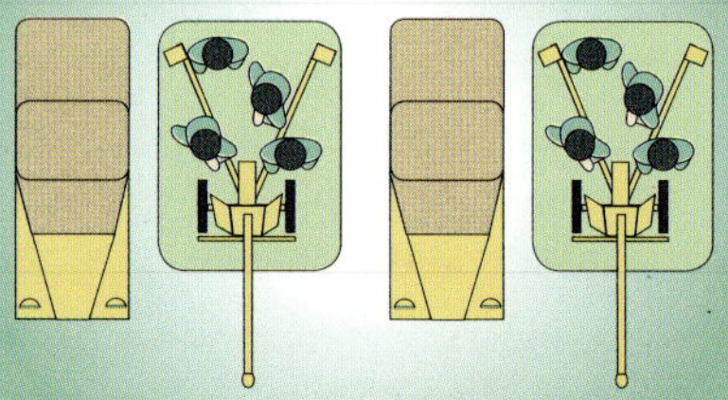

Normandy 1944
Above: an American mixed tank battalion in harbor (US spelling!), with light tanks being re-supplied and two medium tanks being repaired by the unit's armouured recovery vehicle (ARV).
Below: a German infantry battalion on the move in close country.

American Armored Combat Command - June 1944

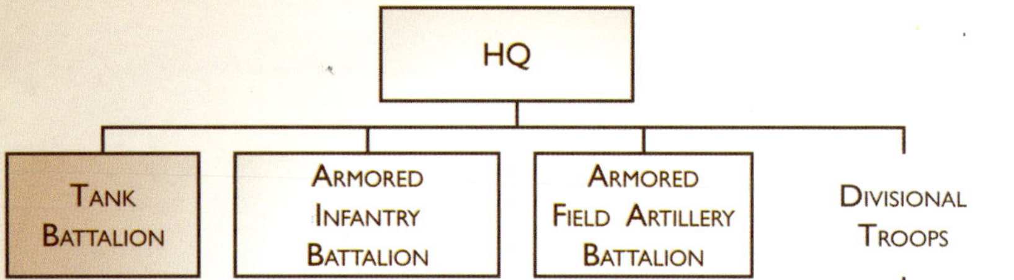

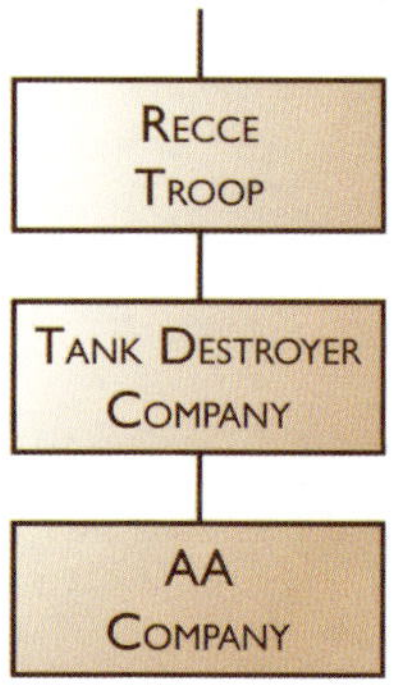

American armoured divisions were organised into three flexible combat commands. The two 'heavy' armoured divisions (2nd and 3rd – the first to fight in Normandy in 1944) had more tanks than the other divisions - six tank battalions (two light, four medium) - rather than three. These were quickly reorganised to create six mixed units. Our combat command has only one of these mixed battalions (smaller than the 'light' division version, which has another company of medium tanks), but a second could be added. The combat command also has a battalion of armoured infantry in M3 half-tracks, well armed with heavy weapons and bazookas, a battalion of self propelled 105mm guns and attached reconnaissance, tank destroyer and anti-aircraft units. After mid-July replace one Sherman 75mm in each company with a 76mm-armed M4A1.

All these US troops are regular.

HQ

6 figures, Jeep, M3 radio half-track
Stuart tank (37mm)

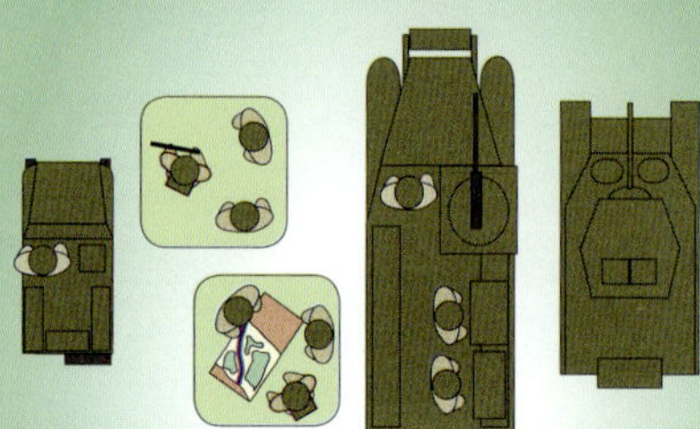

Supply Company
3 x medium lorry

Armored Field Artillery Battalion

3 figures, M3 radio half-track, 3 x M7 SP gun (105mm), OP team, Jeep, Sherman OP tank

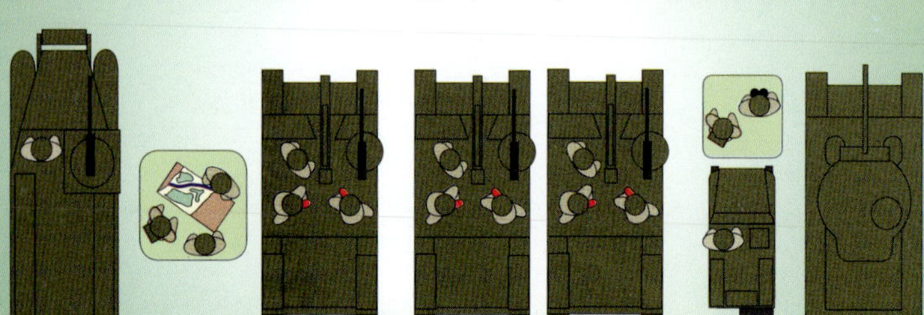

Tank Battalion

HQ
Sherman tank (75mm), Sherman tank (105mm)
Sherman ARV

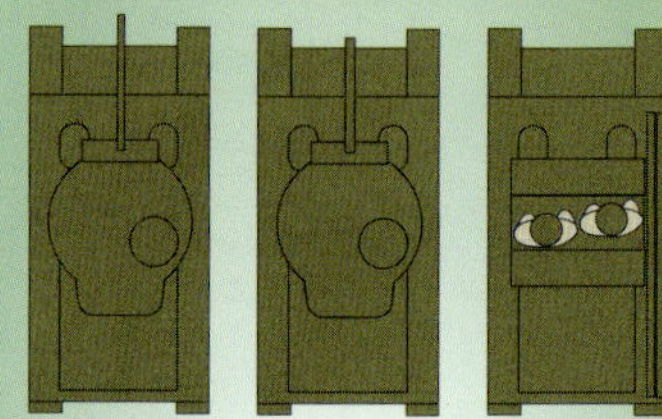

6 figures, 81mm mortar
M4 half-track, Jeep (50cal)

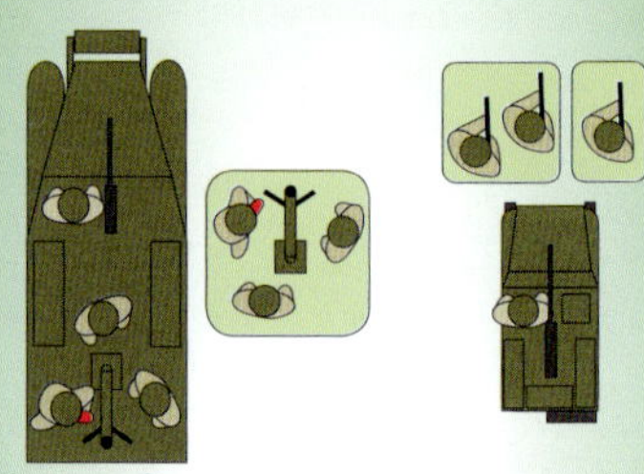

Light Company
3 x Stuart tank (37mm)

Medium Company
3 x Sherman tank (75mm)

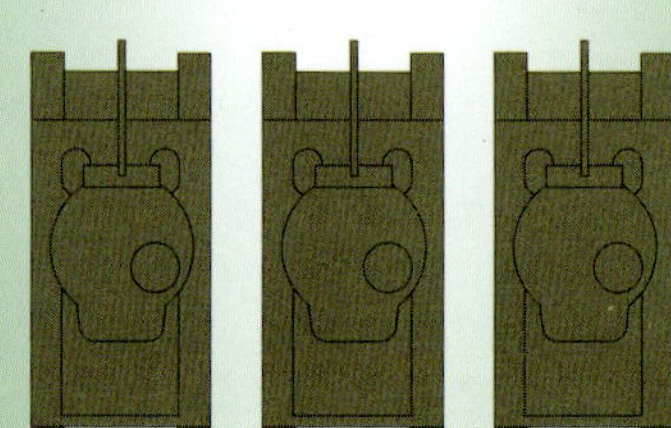

Medium Company
3 x Sherman tank (75mm)

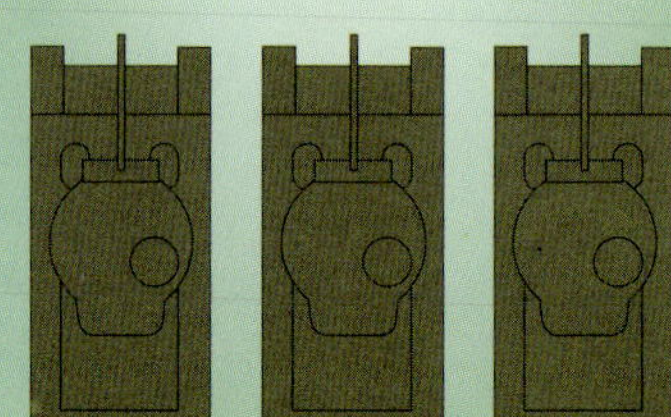

Armored Infantry Battalion (53 figures)

HQ
CO + 6 figs, 60mm mortar, Bazooka
M3 half-track

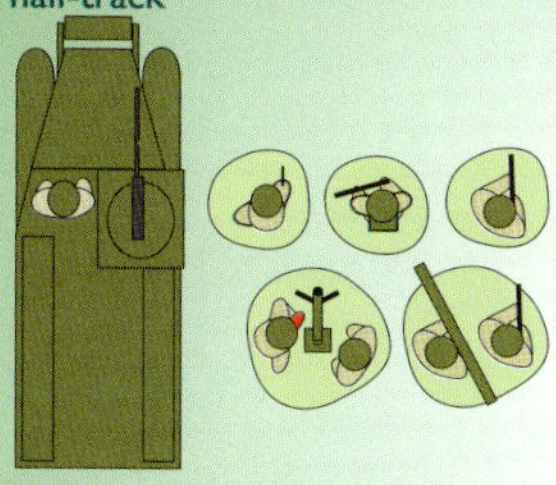

Rifle Company
10 figures, Bazooka, M3 half-track

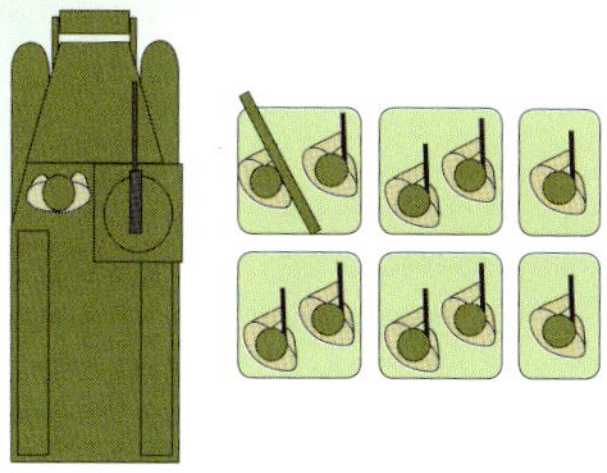

Rifle Company
10 figures, Bazooka, M3 half-track

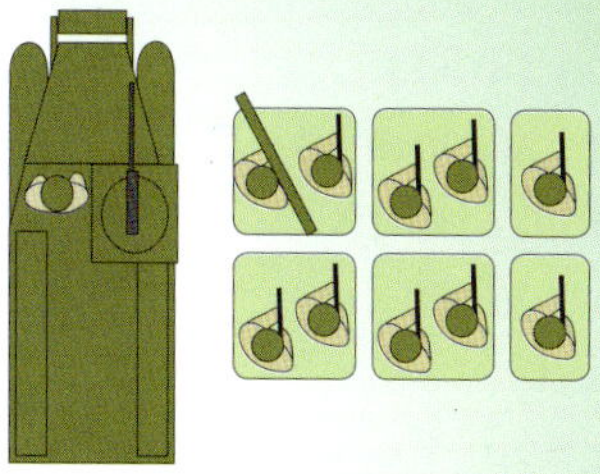

Assault Gun Platoon
M8 HMG (75mm)

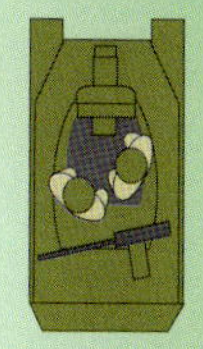

Rifle Company
10 figures, Bazooka, M3 half-track

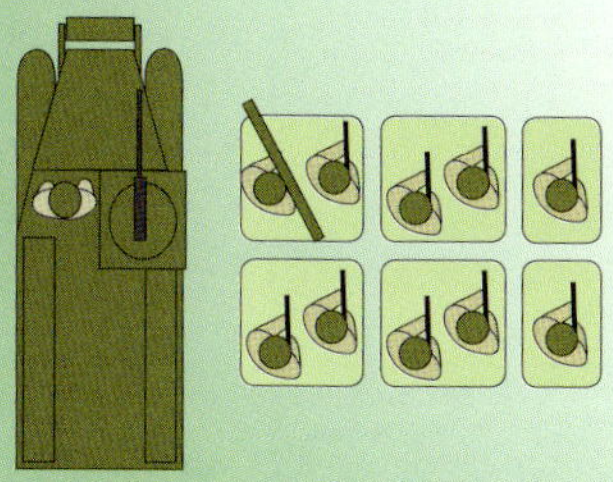

Heavy Weapons Company
6 figures, 50cal HMG, 30cal MMG, Bazooka, M3 half-track, 3 figures, 81mm mortar, M4 half-track
3 figures, 57mm AT gun, M3 half-track

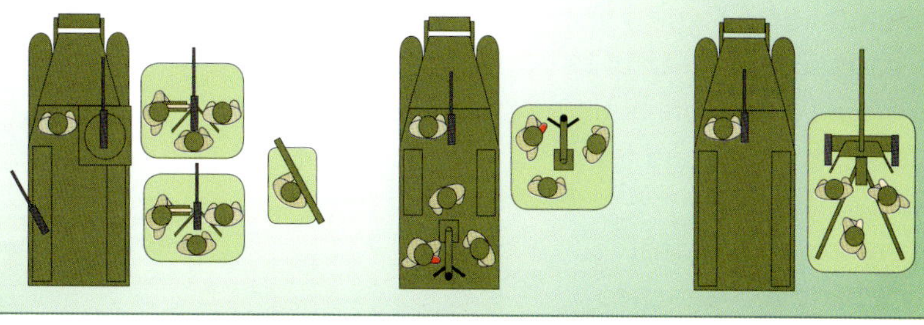

Reconnaissance Troop (13 figures)
CO + 6 figures, 60mm mortar, Bazooka
Jeep (HMG), Jeep, M8 armoured car (37mm)

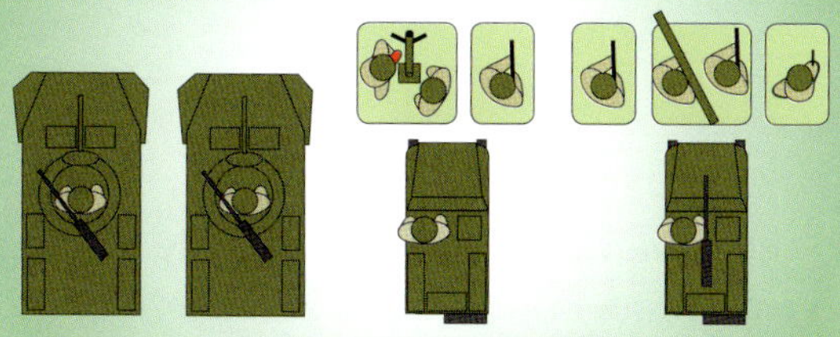

Tank Destroyer Company
2 x M10 tank destroyer (76mm)

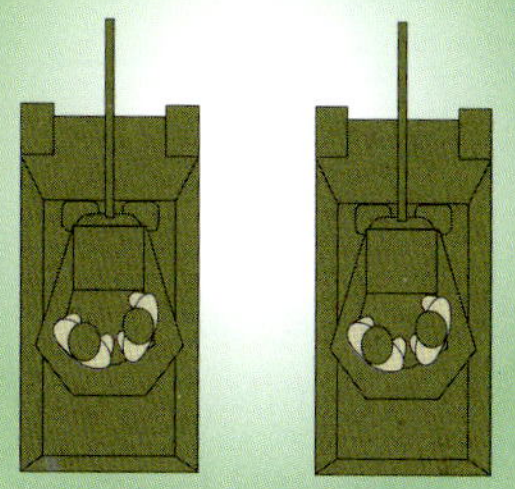

Anti-Aircraft Company
2 x M16 AA half-track, M15 AA half-track

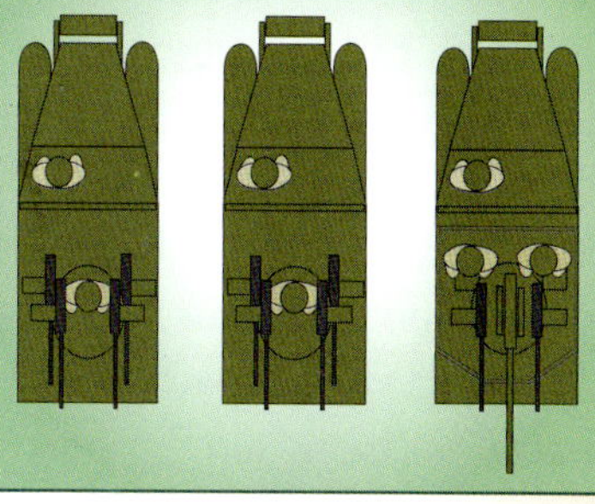

Vehicle and Gun Chart - American Armored Combat Command

Vehicle (Main Weapon)	Speed	Armour	Hull (and Other) Weapons	Turret Weapons
Stuart tank (37mm)	30" / 15"	D	VMG	AT class 4 / HE 6 points table 4 (or VMG)
M8 howitzer motor carriage (75mm)	30" / 15"	D	-	AT class 4 / HE 10 points table 4 (or VMG)
Sherman tank (75mm)	24" / 12"	C	VMG	AT class 3 / HE 10 points table 4 (or VMG)
Sherman tank (76mm)	24" / 12"	C	VMG	AT class 2 / HE 8 points table 4 (or VMG)
Sherman tank (105mm)	24" / 12"	C	VMG	AT class 3 / HE 12 points table 4 (or VMG)
Sherman OP tank	24" / 12"	C	VMG	VMG
M10 tank destroyer (76mm)	24" / 12"	D	-	AT class 2 / HE 8 points table 4 (or VMG)
M8 armoured car (37mm)	30" / 15"	E	-	AT class 4 / HE 6 points table 4 (or VMG)
M7 self-propelled gun (105mm)	24" / 12"	E	AT class 3 / HE 12 points table 4	
Sherman ARV	24" / 12"	C	VMG	
M3 half-track (HMG)	30" / 15"	E	HMG	
M4 half-track (mortar)	30" / 15"	E	81mm mortar or MMG	
M15 AA half-track (37mm)	30" / 15"	E	Multiple autocannon	
M16 AA half-track (quad HMG)	30" / 15"	E	Multiple HMG	
Jeep (HMG)	30" / 18"	F	HMG	
2½ ton truck	24" / 12"	F	-	
Gun				
57mm anti-tank gun	AT class 3 / HE 6 points table 4			

The armoured infantry battalion MG unit has a one figure Bazooka team. Players may add another figure from an MG crew if desired.

American Regimental Combat Team - June 1944

The American infantry division adopted a similar organisation to its armoured equivalent by fielding three regimental combat teams, each with attached supporting units. Our RCT has its own artillery in the form of the cannon company, an anti-tank company (as well as the battalion AT guns) and three battalions, similar to the armoured version, but with less transport and fewer bazookas. The RCT normally has a divisional artillery battalion attached for direct support (see the Specialist Artillery rules), but can call on other battalions through the efficient US artillery control system. Although some infantry divisions had attached self- propelled tank destroyer units, our RCT has a company of towed 3" guns, which were less suited to mobile warfare. As well as reconnaissance and engineer units, our RCT also has a tank company; an independent tank battalion being almost permanently attached to most infantry divisions.

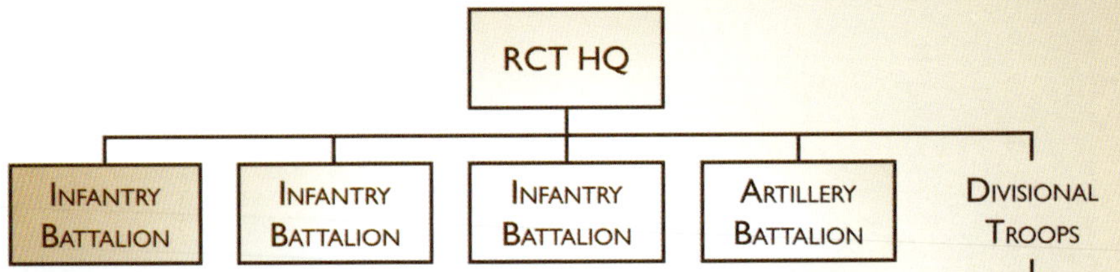

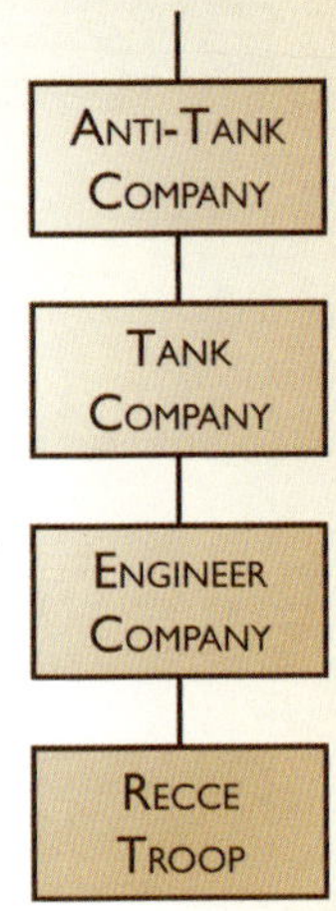

Regimental Combat Team HQ

6 figures, Jeep
2½ ton radio truck

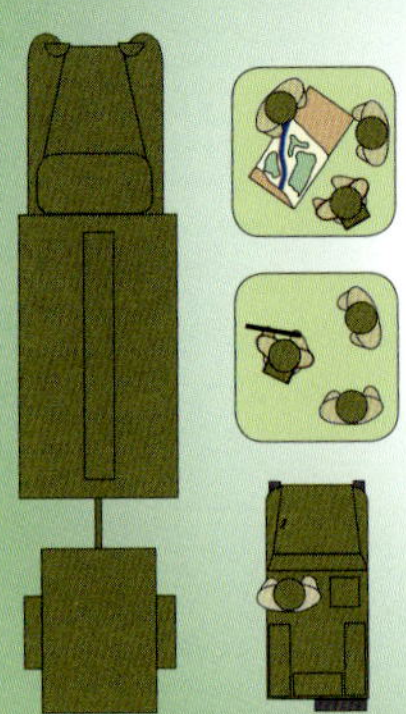

Cannon Company
3 figures, 105mm 'short' howitzer, 1½ ton truck

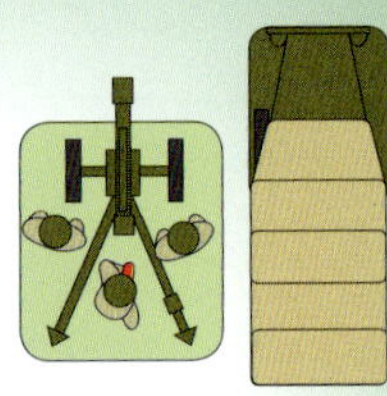

Anti-Tank Company
3 figures, 57mm anti-tank gun, 1½ ton truck

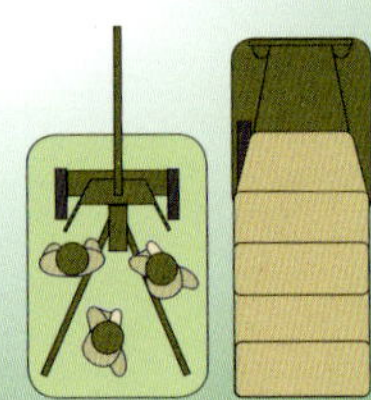

All these US troops are regular.

Infantry Battalion (48 figures)

HQ
CO + 5 figures, 60mm mortar, ¾ ton truck

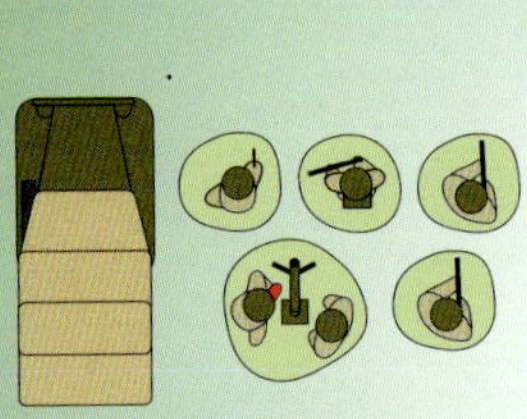

Rifle Company
10 figures, Bazooka, 2½ ton truck

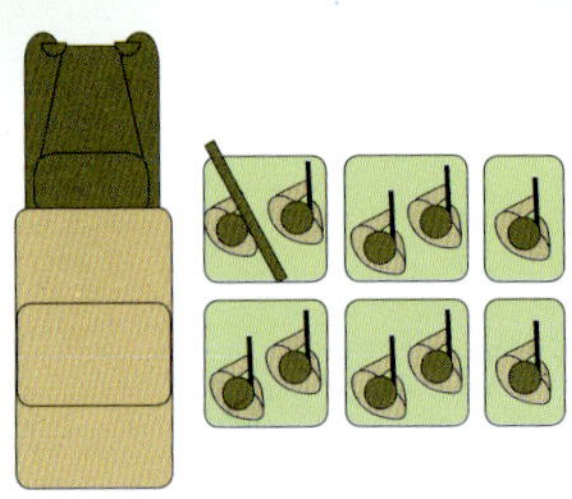

Rifle Company
10 figures, Bazooka, 2½ ton truck

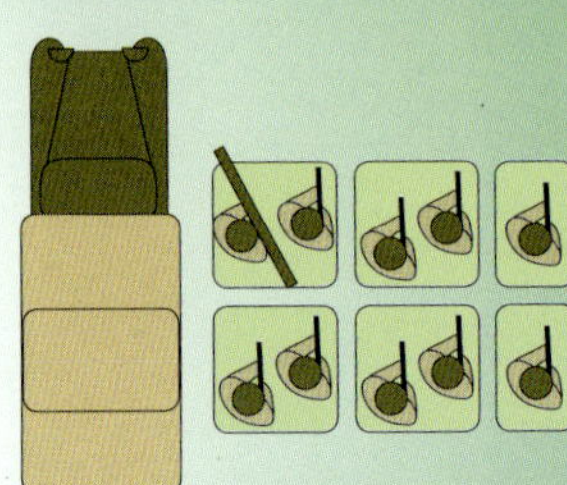

Rifle Company
10 figures, Bazooka, 2½ ton truck

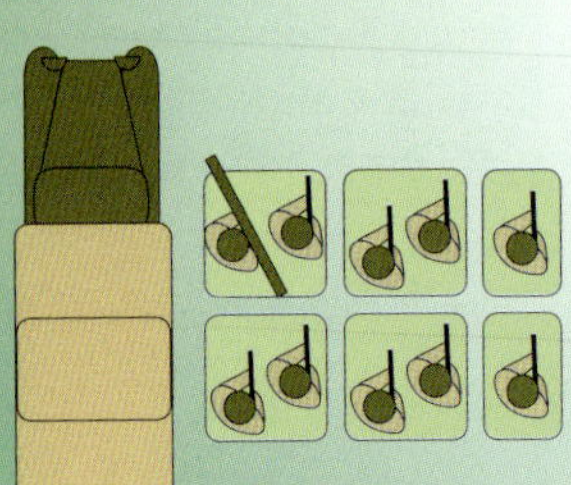

Heavy Weapons Company
3 figures, 50cal HMG, Jeep, 3 figures, 30cal MMG, Jeep,
3 figures, 81mm mortar, Jeep and trailer, 3 figures, 57mm AT gun, 1½ ton truck

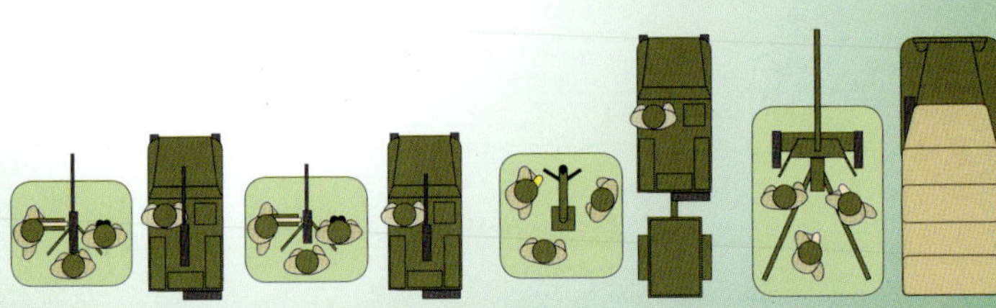

Artillery Battalion

Battery
14 figures, 3 x 105mm howitzer, OP team, 3 x 2½ ton prime mover, 2 x Jeep

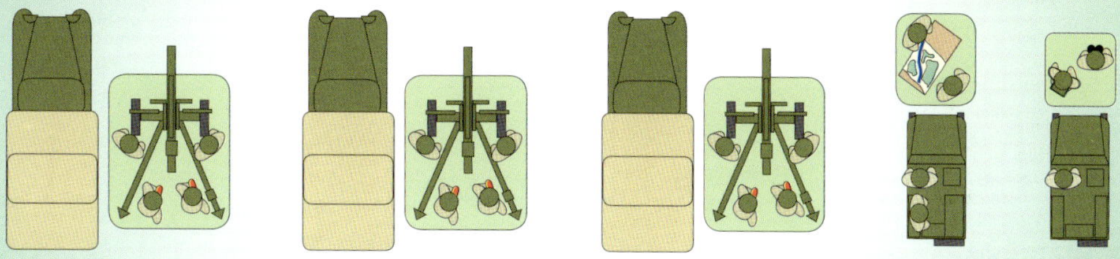

Anti-Tank Company
8 figures, 2 x 3" anti-tank gun, 2 x M3 half-track

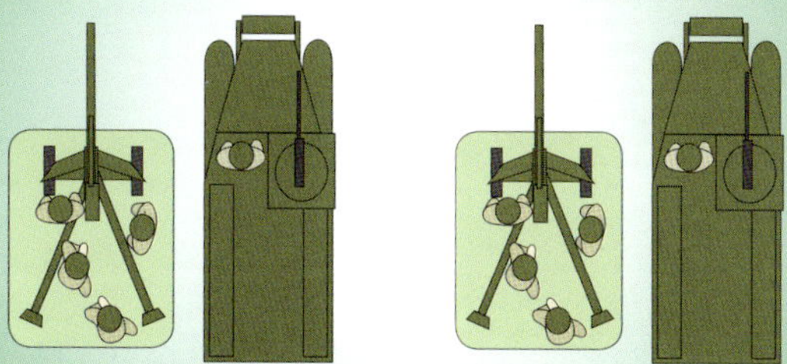

Tank Company
3 x Sherman (75mm)

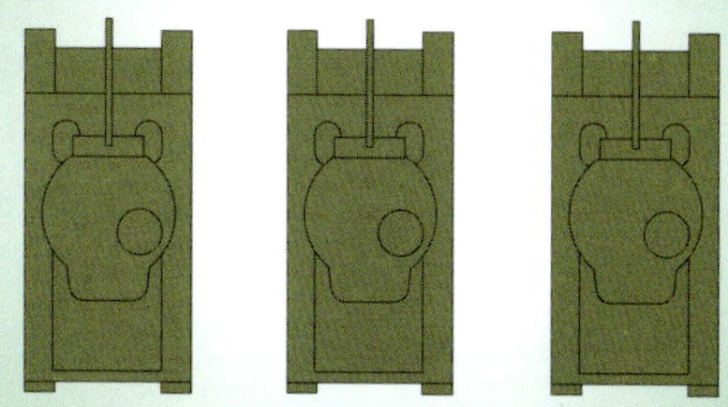

Engineer Company
10 figures, Bazooka, 2½ ton tipper truck

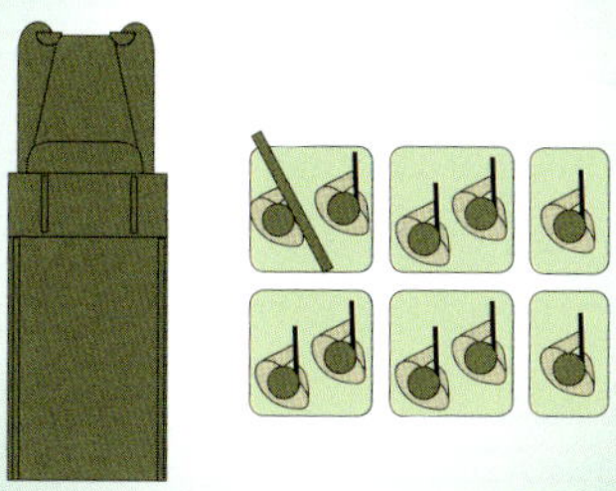

Reconnaissance Troop (13 figures)
CO + 6 figures, 60mm mortar, Bazooka
Jeep (HMG), Jeep, M8 armoured car (37mm)

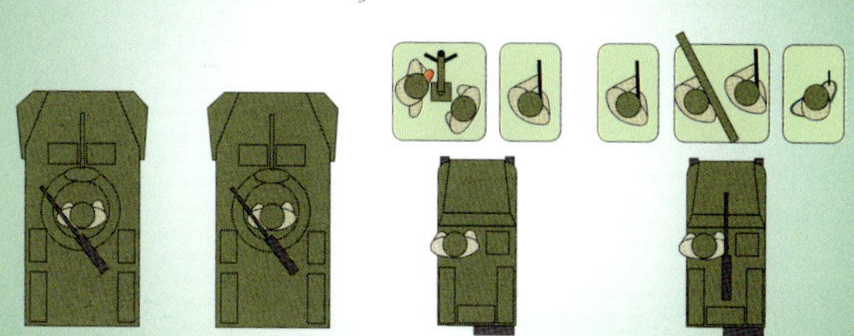

Vehicle and Gun Chart - American Regimental Combat Team

Vehicle (Main Weapon)	Speed	Armour	Hull (and Other) Weapons	Turret Weapons
Sherman tank (75mm)	24" / 12"	C	VMG	AT class 3 / HE 10 points table 4 (or VMG)
M8 armoured car (37mm)	30" / 15"	E	-	AT class 4 / HE 6 points table 4 (or VMG)
M3 half-track (HMG)*	30" / 15"	E	HMG	
Jeep (HMG)	30" / 18"	F	HMG	
¾ ton / 1½ ton* truck	30" / 15"	F	-	
2½ ton / radio / tipper truck / prime mover	24" / 12"	F	-	
Gun				
57mm anti-tank gun	AT class 3 / HE 6 points table 4			
3" anti-tank gun	AT class 2 / HE 8 points table 4			
105mm howitzer	AT class 3 / HE 12 points table 4			
105mm 'short' howitzer	AT class 4 / HE 12 points table 4			

* Maximum prime mover movement when towing a gun is 24" / 12".

Allied armies in Normandy had the advantage of massive air superiority and ground attack aircraft such as the American Thunderbolt played a major part in defeating the Germans. All Rapid Fire! aspects of aircraft are covered in the Advanced rules section which starts at page 94.

Russian Tank Brigade Forward Detachment - Summer 1944

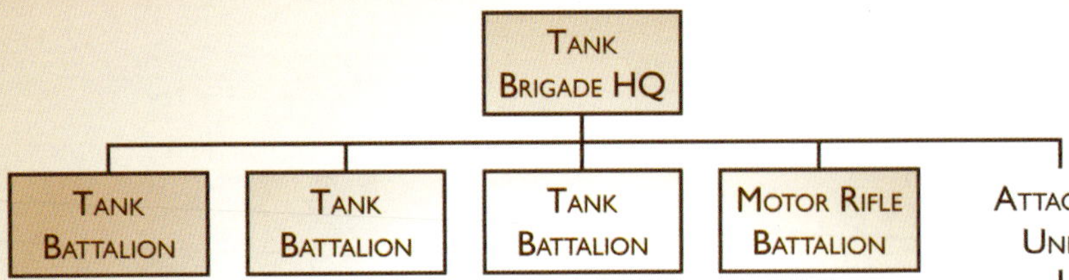

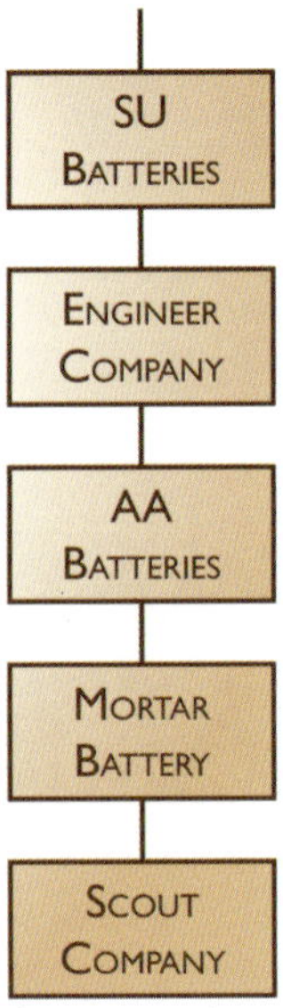

The Russian tank brigade developed from an emergency organisation in 1941 to the building blocks of the new tank corps of 1942 onwards. In 1944, as Russian units spent more and more time advancing against weakening German opposition, the need for the tank corps to quickly seize strategic points (such as bridges) became more urgent. The concept of the forward detachment was unique to the Russian army and our force is a complete tank brigade, with supporting units that is intended to speed ahead of the main force and capture key objectives. It has been equipped with the latest 85mm-armed T34s, but has a very small infantry battalion that mainly rides into battle on the tanks, although its SMG armament provides devastating short-range fire. All the detachment's artillery support is intended for direct fire, with no OPs, and it has a strong AA company to counter air attacks and defend captured localities. As river crossings are the detachment's most common objective the scout company is equipped with lend-lease amphibious Jeeps, to enable it to cross and establish bridgeheads, attack bridge defenders from the rear and ferry over the SMG companies.

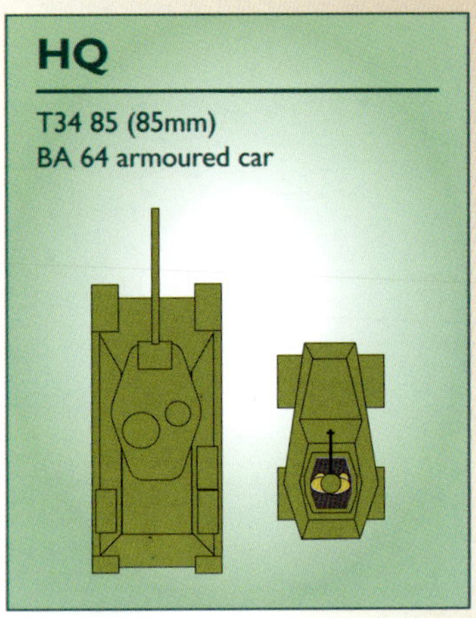

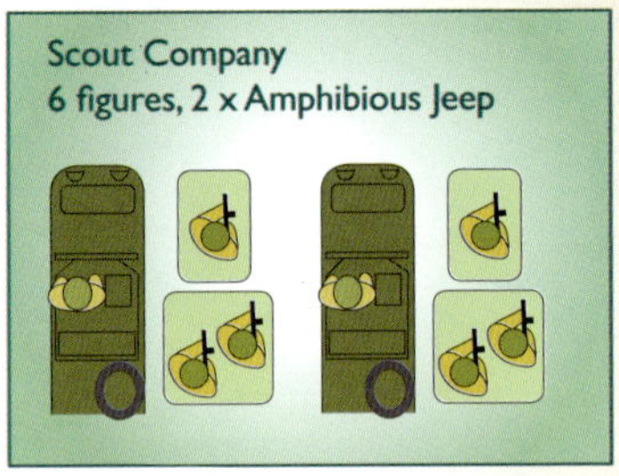

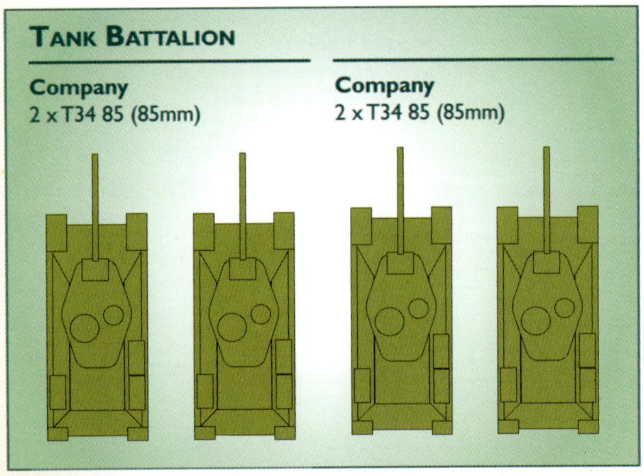

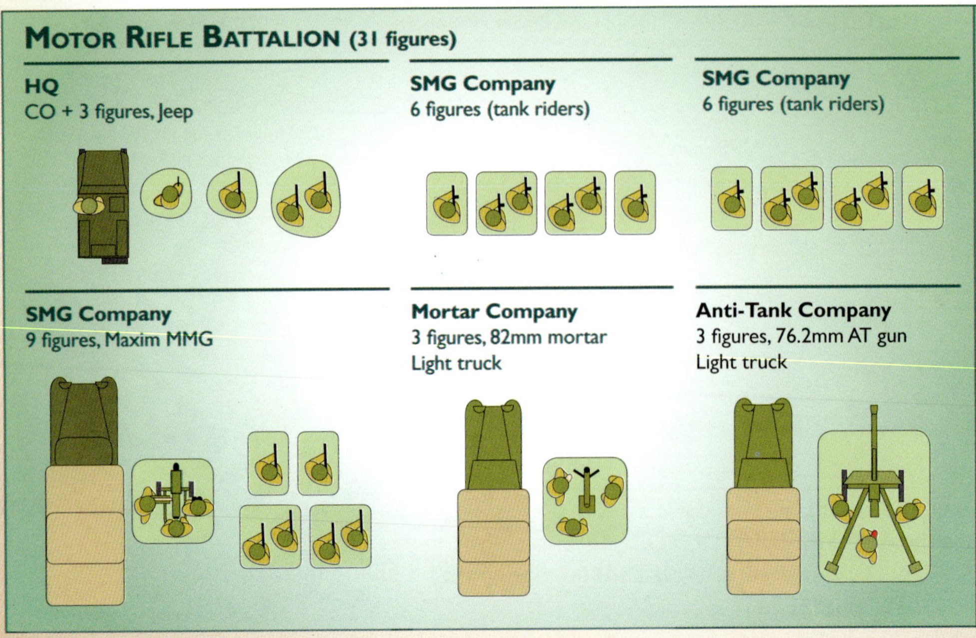

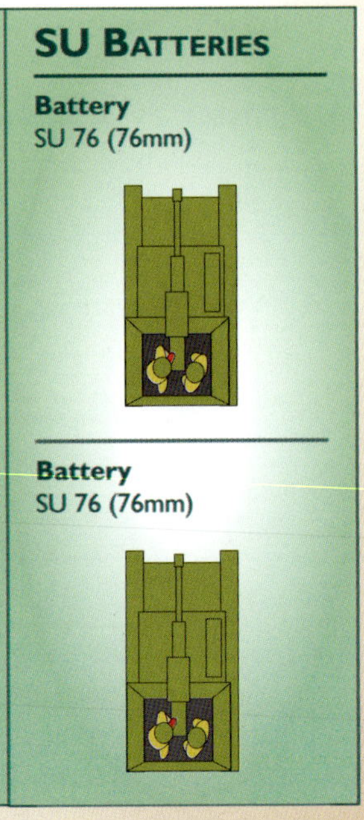

Anti-Aircraft Batteries

Battery
3 figures, 37mm AA gun
Light truck

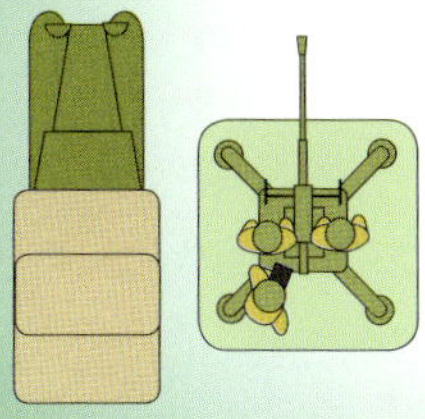

Battery
Zis AA half-track (37mm)

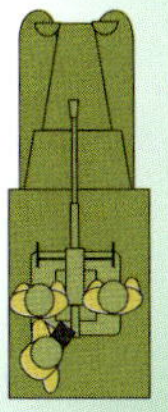

Engineer Company
7 figures, flamethrower
Medium truck

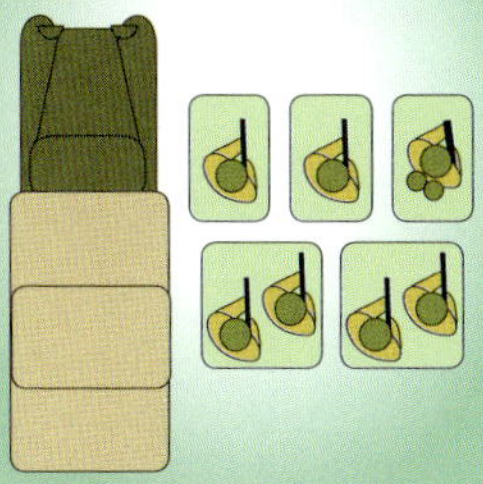

Mortar Battery
BM 13 Katyusha
multiple rocket launcher

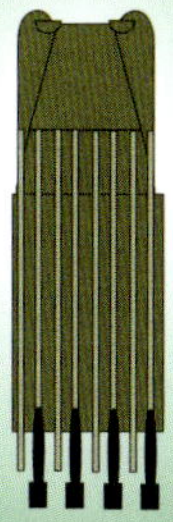

Vehicle and Gun Chart - Russian Tank Brigade Forward Detachment

Vehicle (Main Weapon)	Speed	Armour	Hull (and Other) Weapons	Turret Weapons
T34 / 85 tank (85mm)	30" / 15"	B	VMG	AT class 2 / HE 10 points table 4 (or VMG)
BA 64 armoured car (HMG)	30" / 15"	E	-	HMG
SU 76 assault gun (76mm)	24" / 12"	E	AT class 3 / HE 10 points table 4	
BM 13 Katyusha multiple rocket launcher	30" / 15"	F	Multiple rocket launcher	
Zis AA half-track	30" / 15"	F	37mm autocannon	
Jeep / Amphibious Jeep*	30" / 15"	F	-	
Light truck**	30" / 15"	F	-	
Medium truck	24" / 12"	F	-	
Gun				
76.2mm anti-tank gun	AT class 2 / HE 10 points table 4			
37mm anti-aircraft gun	37mm autocannon			

* An Amphibious Jeep can cross water obstacles at 4" per turn.

** Maximum prime mover movement when towing a gun is 24" / 12"..

All Russian troops are regular.

Western Russia 1944
Above: a scout and tank company approach the River Dvina.

Left: tank riders improvise a river crossing covered by SU 76s.

Top: a deployed British field artillery regiment.
Middle: a German horse drawn light artillery battalion on the move.
Left: a German 20mm AA company (14th).

Order of Play

Rapid Fire! uses an alternate play system, with side A taking their turn to carry out all their actions first, followed by side B. Each pair of turns (A + B) makes up one move of the game.

During each side's turn, follow the **Order of Play** below, omitting any of the nine stages that are not applicable. For example, a **Morale** test will not be necessary on move 1 and a **Heroic Action** test may only arise once or twice in a game. The only exception to the order of play is automatic **Observation**, which can take place at any time during the turn.

1	**Morale**	Take all necessary morale tests and carry out compulsory reactions.
2	**Observation**	Take all necessary observation tests.
3	**Smoke**	Fire smoke rounds.
4	**Heroic Action**	Attempt tests for heroic actions.
5	**Movement**	Move each element now <u>or</u> at the end of the turn, but not both. Movement made now may end in close assault.
6	**Reserved Fire**	Allow your opponent to declare and carry out any reserved fire. In specific situations this may interupt movement.
7	**Close Assault**	Resolve any close assaults.
8	**Firing**	Declare all elements that you wish to fire and clearly identify their targets. Carry out firing tests.
9	**Movement**	Move any units that have not yet moved this turn*. Movement made now may not end in close assault.

* Some weapons (such as battalion mortars, MMGs and artillery batteries) cannot be moved and fire in the same turn. See **rule 9.2**.

The layout of the **Basic Rules** follows these nine stages and each rule or explanation is numbered for ease of reference.

The **Advanced Rules** are also linked to this order of play. Players should agree which of the advanced rules (if any) are to be use before the game begins.

Morale

When units suffer losses or face particular threats in combat they can react by becoming less reliable. In extreme cases they may retreat, run away (rout), or surrender. Some units are less likely and some more likely to react poorly to these situations, being regarded, respectively, as having higher or lower **Morale** than normal.

1.1 Units with higher morale are classed as elite. Units with lower morale are classed as poor and can include any troops who are poorly trained, led or equipped. All other units (the majority) are classed as regular.

1.2 The morale classification of units should be agreed before the game and elite unit status should only be granted where there is historical justification (for example: commandos are usually regarded as elite).

1.3 Morale tests are carried out at the beginning of each player's turn by:

- Infantry or cavalry units of battalion or equivalent size that have suffered a total of 20 casualties (or 50% of their original strength if they started the game with less than 40 figures) after losses during the enemy's last turn.
- Independent armoured fighting vehicles (AFVs) that have suffered heavy damage during the enemy's last turn.
- Armoured units in which an AFV has suffered heavy damage or has been destroyed during the enemy's last turn.
- Any infantry or cavalry unit on foot, horseback or in vehicles, and any deployed guns that have been targeted by indirect artillery fire (IDF) from 80mm or larger calibre guns (not mortars), or which have been attacked by aircraft during the enemy's last turn.
- Infantry / cavalry in the open that can see an enemy AFV within 12" of their position at the beginning of their turn.

Infantry and Cavalry Units

1.4 When an infantry or cavalry unit of battalion or equivalent size has lost 20 figures (or 50% casualties if it started the game with less than 40 figures) it must take a morale test. If it then suffers further losses at any future point in the game it must take a morale test in the turn immediately afterwards.

1.5 To decide what the reaction of the unit is to these losses throw a D6 and add (+) or subtract (-) any relevant modifiers. The resulting number dictates how the unit must react. **Table 1** gives dice modifiers and reactions.

To ease calculation of figure losses, keep each unit's casualties off-table, either in a small tray, or in a section of a divided container.

Playing Tip

Table 1: Infantry and Cavalry Morale

Dice Score Modifiers	
-1	If the unit's commanding officer (CO) is one of the casualties.
-1	If the unit is classed as poor.
-1	If the unit is unsupported with no element* (i.e. part) of a friendly unit within 12" of any of its own elements. This does not apply if the unit is elite and normally moves and fights independently (such as paratroops, commandos, LRDG etc).
-1	If any part of the unit has been targeted by indirect fire from guns (but not mortars) of 80mm calibre or greater.
-1	For each consecutive turn that a morale test is necessary**.
+1	If the unit is classed as elite.
+1	If the majority (more than 50%) of the surviving figures in the unit are in hard cover.
+2	If all the surviving figures in the unit are occupying prepared defences.

* For morale purposes an element does not mean an uncrewed weapon or a transport vehicle.

** For example, a unit that had to take a morale test last move is also forced to take one this move. This is the second consecutive turn in which a morale test has been necessary so the dice throw is modified by -1. This will become -2 for the third consecutive test, -3 for the fourth and so on. If no casualties are suffered in a turn then no morale test is taken. If further casualties result in the need for a morale test the whole process has to start again.

Results and Reactions		
4 or more	NO EFFECT	
3	NO OFFENSIVE ACTION	The unit can fire, but it may not advance towards the enemy. If any element is caught in the open it must withdraw a full move towards the nearest cover occupied by, or closest to, friendly elements.
2	PINNED DOWN	Units in cover may not move but may fire at reduced effect. Deduct 2 from a D6 throw for small arms fire effect. Mortars may fire one round only. AT weapons deduct 2 from their D6 throw when testing for a hit. Units in the open may not fire and must withdraw a full move towards the nearest cover occupied by, or closest to friendly elements.
1 or less	SURRENDER or ROUT	The unit surrenders if it is cut off from friendly troops. Otherwise, it routs off table (see **rule 1.14**).

A German infantry element surrenders to British troops.

Independent AFVs and Armoured Units

1.6 A morale test must be taken:

- When an independent AFV has suffered heavy damage, or
- When an AFV in an armoured unit (normally a company or squadron of two to four vehicles) has suffered heavy damage, or has been destroyed, or
- When either an already heavily damaged independent AFV, or an already heavily damaged AFV in an armoured unit, suffers light damage.

1.7 To decide what the reaction of the unit is to this damage (or these losses) throw a D6 and add (+) or subtract (-) any relevant modifiers. The resulting number dictates how the unit must react. **Table 2** gives dice modifiers and reactions.

1.8 If a morale test results in an AFV immobilised by heavy damage in the previous turn having to retire, it must be abandoned by its crew.

1.9 AFVs have a nominal crew of three figures. Units that include independent AFVs or squadrons / companies of AFVs, such as reconnaissance battalions, use this number of crew-per-vehicle to determine the figure strength of the 'parent' unit and to work out when 50% or 20 casualties have been lost.

Example of AFV Morale

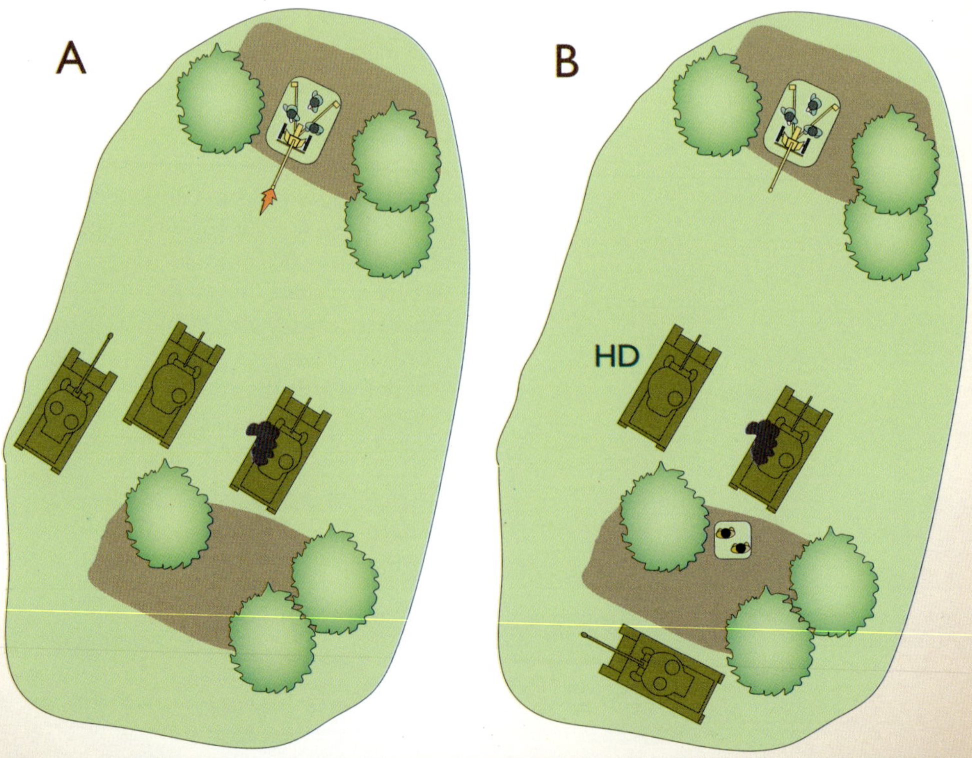

A

During its last turn a British tank squadron advancing towards a wood lost one 75mm Sherman to the reserved fire of a German 75mm AT gun. The British fired back and missed. In the German turn the surviving Sherman 75mm is fired at by the same gun and is heavily damaged.

B

At the beginning of the next British turn a D6 is thrown to test the squadron's morale. A 4 is thrown. 2 is deducted (-1 for both the destroyed and the heavily damaged tanks). The result is 'no offensive action' and the crew of the Sherman 75mm bail out and take cover in the nearby wood. The Firefly withdraws behind the wood.

TABLE 2: AFV MORALE

Dice Score Modifiers		
-1	For each heavily damaged or destroyed AFV in the unit at the time of the test.	
-1	If the independent AFV or any vehicle in the unit is within 12" of an observed enemy element occupying hard cover.	
-1	For an independent AFV that has suffered heavy damage.	
-1	If the independent AFV or unit is classed as poor.	
+1	If the independent AFV or unit is classed as elite.	
Results and Reactions		
3 or more	NO EFFECT	
2 or 1	NO OFFENSIVE ACTION	The AFV (or AFVs if in a unit) can still fire unless heavily damaged in the previous turn, but must not advance towards the enemy. If any vehicle is caught in the open it must withdraw a full move towards the nearest cover occupied by, or closest to, friendly elements*. If already in cover it may not move.
0	RETIRE TWO MOVES	This movement is carried out immediately. Affected vehicles must move away from the enemy. They cannot fire and and do not take part in the game during this and the following turn* (see **rule 1.14**).
-1 or less	SURRENDER or ROUT	The independent AFV (or AFV unit) surrenders if it is cut off from friendly troops, otherwise it routs off table (see **rule 1.14**).

* If an AFV is immobilised by heavy damage, mark its 'retire' result by placing a base of two tank crew figures 6" away from the vehicle. These figures cannot fire, or be fired at, or re-crew another AFV. They are removed in the first move that is free of a morale test. The AFV is then re-activated in the following friendly turn.

Below: a two figure AFV morale marker.

Indirect Artillery Fire and Attacks by Aircraft

1.10 When an infantry unit (on foot or in vehicles), cavalry unit, or deployed gun has been targeted by indirect artillery fire (IDF) by guns of over 80mm, or has been fired at or bombed by aircraft, it must test to determine if it is 'suppressed'. Note that AFVs and troops in prepared defensive positions (such as trenches, bunkers etc) do not have to take this morale test.

1.11 To decide what the reaction of the unit is to this fire, or bombing, throw a D6 and add (+) or subtract (-) any relevant modifiers. The resulting number dictates how the unit must react. **Table 3** gives dice modifiers and reactions.

1.12 If IDF or air attack causes casualties to an infantry or cavalry unit that bring total losses to 50% or 20 figures, then it must first take a morale test under **rule 1.4**. If the unit passes (i.e. a result of 4 or more) it must then take the IDF / air attack test (**rule 1.11**). If the unit fails, carry out the reactions specified in **Table 1** and ignore the IDF / air attack test.

TABLE 3: INDIRECT ARTILLERY FIRE (IDF) AND ATTACKS BY AIRCRAFT

Dice Score Modifiers		
-1	If the unit is classed as poor.	
+1	If the unit is classed as elite.	
+1	If the element is infantry, or gunners, occupying hard cover.	
Results and Reactions		
4 or more	NO EFFECT	
3 or less	SUPPRESSED	The unit may not move or fire during this turn. Infantry, cavalry and gunners in the open lay down and are treated as if in soft cover. Any figures being carried in soft-skin vehicles debuss and lay down beside their transport.

In situations when units are partly in and out of prepared positions only those with a majority of figures in the open need to test. Left: no test is required for this infantry company.

Infantry and Cavalry in the Open

1.13 A morale test is necessary if any unsupported infantry or cavalry unit (i.e. with no friendly AFVs within 12") is in the open and an enemy AFV has moved to within 12" of its position. Throw a D6 and add (+) or subtract (-) any relevant modifiers. The resulting number will decide whether or not the infantry have to move away from the enemy AFV. **Table 4** gives dice modifiers and reactions.

Table 4: Infantry and Cavalry in the Open

Dice Score Modifiers		
-1	If the unit is classed as poor.	
+1	If the unit is classed as elite.	
Results and Reactions		
4 or more	NO EFFECT	
3 or less	RETIRE	The infantry or cavalry immediately take a full move away from the enemy AFV and towards its own forces. Infantry may enter a building to avoid the AFV (see **rule 1.14**).

When suppressed troops (like these German panzer grenadiers) have to abandon vehicles they 'lay down' beside them rather than moving 3".

Retiring, Routing and Surrender

1.14 Always carried out immediately, a 'retire' result means a unit must move its maximum movement distance (double if retiring two moves) away from the enemy and towards a friendly unit, if any are still on the table. No deductions to movement distances are made for terrain and continuous and linear obstacles. Infantry units retiring from an AFV may avoid it by entering buildings.

Units that rout (run away) are treated as retiring for two full moves, every move, until they reach the edge of the table. They are then removed from the game. They will use roads wherever possible and this may block or delay advancing friendly units. Units forced to surrender are immediately removed from the table.

Chance Observation tests must be taken at this point, but elements may 'spot' and react to enemy units within **Automatic Observation** ranges throughout the turn.

2.1 A target can normally only be fired at if it has been observed. Speculative firing (e.g. 'I'll fire at that hedgerow, there may be troops behind it') is not permitted, unless firing smoke, targeting buildings or bridges, using IDF or specialist artillery rules (see **advanced rules 14.2 and 14.7**). See **rule 8.3** for a full definition of a target.

2.2 Infantry, gunners and all reconnaissance vehicles are assumed to have 360° (all round) visibility, but tanks and assault guns may only observe elements in front of a line drawn laterally through the rear of the vehicle's hull (see below). The position of model crew figures on individual vehicles makes no difference to these rules.

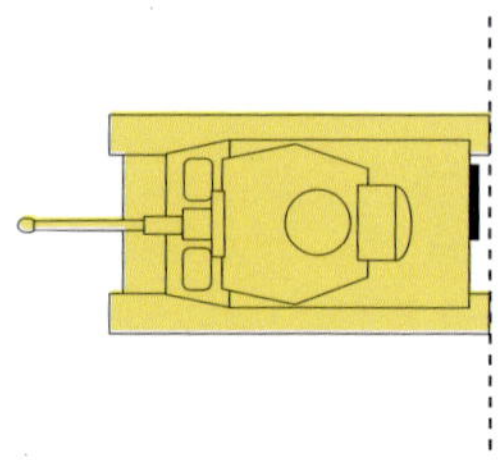

'Blind area'

> Not allowing speculative fire may appear slightly unrealistic, but it prevents a player unfairly targeting unobserved forces simply because he can see his opponent's models on the tabletop.
>
> **Design Note**

2.3 Whether or not a target is in cover, in the open, or has fired in its previous turn, and how far away it is from the observing element determines if:

- Observation of the target is automatic, or
- There is a chance of observing the target by taking a test, or
- The target is out of observation range.

Chance Observation Tests

2.4 A D6 throw decides the result of a chance test. A 5 or 6 is needed to observe an target in soft cover. A 6 is needed to observe any target in hard cover and infantry that have not yet opened fire who are concealed in any intact buildings. If all figures and models in cover are positioned on the table the observing player is free to select which target or targets within chance observation test range they wish to test for. If a test fails, the element must continue to be regarded as unobserved. Players that prefer to use markers to represent units in cover should refer to **rule 2.13**.

2.5 **Table 5** gives automatic and chance observation test ranges. These ranges may be modified by specialist troops as follows:

- Designated reconnaissance units add 6" to the normal observation ranges and artillery and 'airlink' observers (OPs) add 12" (see **Table 5**).

2.6 Only one chance observation test is permitted per target in each turn. If the test succeeds the target is visible to all other observing units within the chance observation test range.

A British para recce squadron (as above) has an observation advantage over normal troops.

Table 5: Automatic and Chance Observation Test Ranges

Target	Automatic Observation (up to)				Chance Observation Test (5 or 6 if in soft cover, 6 if in hard cover)		
	Fired last move?	Normal	Recce	OP	Normal	Recce	OP
Infantry, support weapons* and deployed AT guns up to 75mm in the open	No	36"	42"	48"		-	
	Yes	48"	54"	60"			
Other targets** in the open	No	48"	54"	60"			
	Yes	60"	66"	72"			
Infantry, support weapons* and deployed AT guns up to 75mm in cover*	No	6"	12"	18"	6" - 12"	12" - 18"	18" - 24"
	Yes	36"	42"	48"		-	
Other targets** in cover	No	12"	18"	24"	12" - 24"	18" - 30"	24" - 36"
	Yes	48"	54"	60"		-	
Infantry and support weapons* in intact buildings***	No	No automatic observation			0 - 12"	0 - 18"	0 - 24"
	Yes	36"	42"	48"			
Infantry and support weapons* in ruins***	No	6"	12"	48"	6" - 12"	12" - 18"	18" - 24"
	Yes	36"	42"	48"			
Dummy markers in cover		12"	18"	24"	12" - 24"	18" - 30"	24" - 36"

* Support weapons include all crew-served machine guns, mortars and anti-tank guns up to 75mm.

** Other tagets include artillery pieces, anti tank guns over 75mm and all vehicles.

*** If any of these have been observed moving into an intact or ruined building, the building may be targeted. AT guns may not be deployed in intact buildings.

Cover from Observation

2.7 Observation is blocked at normal table top level by any hills, woods, buildings and smoke screens between an observing element and their target. Due to the exaggerated height of the models in use compared to the figure and vehicle scale, figures and vehicles cannot block observation.

2.8 Some hill tops, church towers and very tall buildings will allow observing elements to see over other buildings and woods (but not other hills and smoke) between them and their target, whatever the height of the model(s) on the wargames table. The relevant areas of hills and buildings which provide this height advantage must be made known to all players before the game begins. If they are not, no height advantage may be claimed during the game.

2.9 Elements in the open within 4" of the far side of a building or wood, and which are no wider than the obstacle, remain concealed from elements observing from any of the higher positions listed in **rule 2.8**. See the example below.

OBSERVATION FROM A HILL

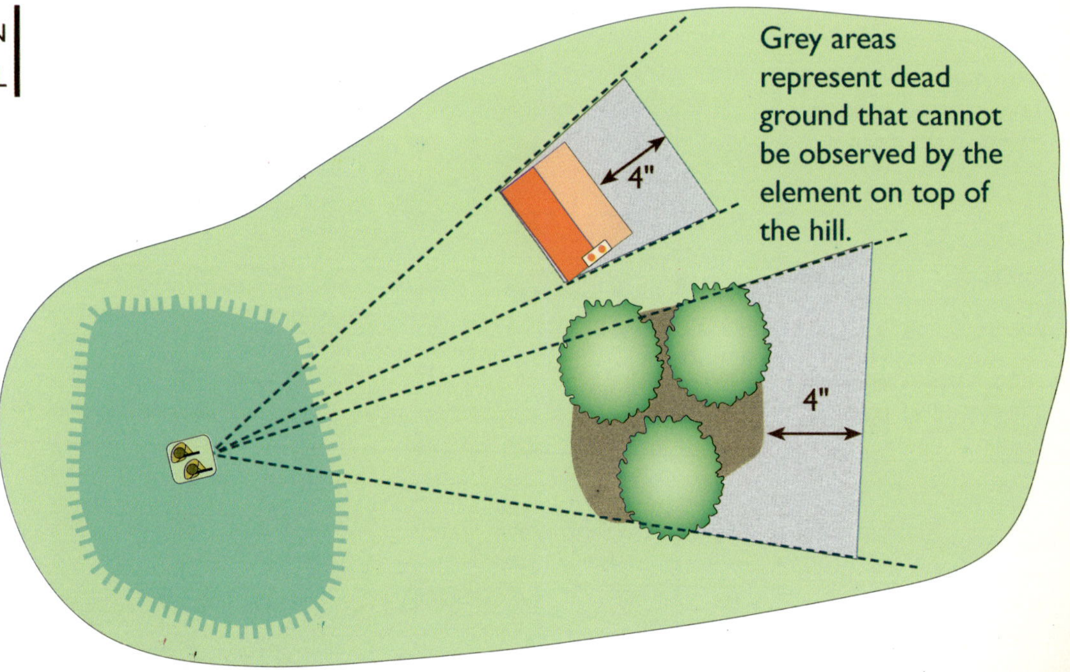

2.10 A smoke marker or smoke screen (see **rules in section 3**) within 2" of an element and between it and the observing element, will conceal it from all ground observation.

2.11 The maximum observation range inside woods, including from the inside out to the edge of the wood and from the edge into the wood is 3". All units inside and beyond this distance from the edge of the wood are completely hidden from enemy forces outside. Any units in woods within 3" of the edge may also see out and observe at normal ranges and are considered to be 'in cover' from any enemy trying to observe them from outside (where normal ranges apply).

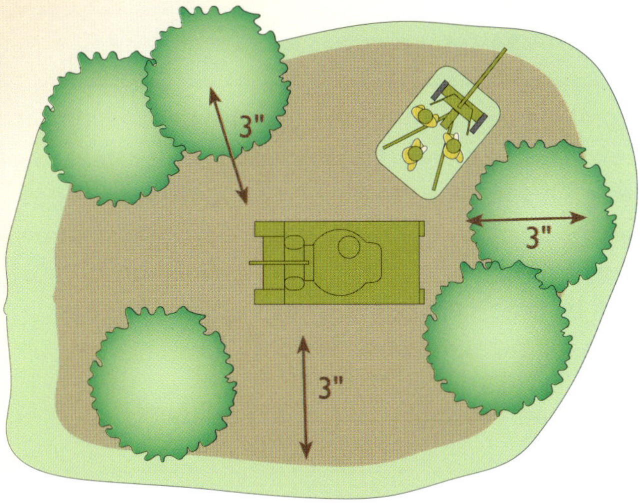

The Sherman tank cannot be seen by any enemy element outside the wood. Enemy elements outside the wood within 12" of the 6 pdr AT gun (which has not fired) can take a chance test of observing it and can see it automatically if they move within 6".

Woods and Cover

Always define the edges of woods and the heights of hedges and walls before the game to avoid disputes later.

Playing Tip

2.12 Unless specified before a game, walls and hedges may only provide cover to infantry, support weapons and deployed anti-tank guns up to 75mm. Vehicles behind and in contact with hedges are classed as partially concealed targets when assessing hits from anti-tank fire. To claim cover, an element's base must be in contact with the hedge or wall.

Right: a feature of the fighting in Normandy was the high-banked 'bocage' hedgerow. With agreement, this can be classed as hard cover and high enough to conceal vehicles.

Hedges, Walls and Cover

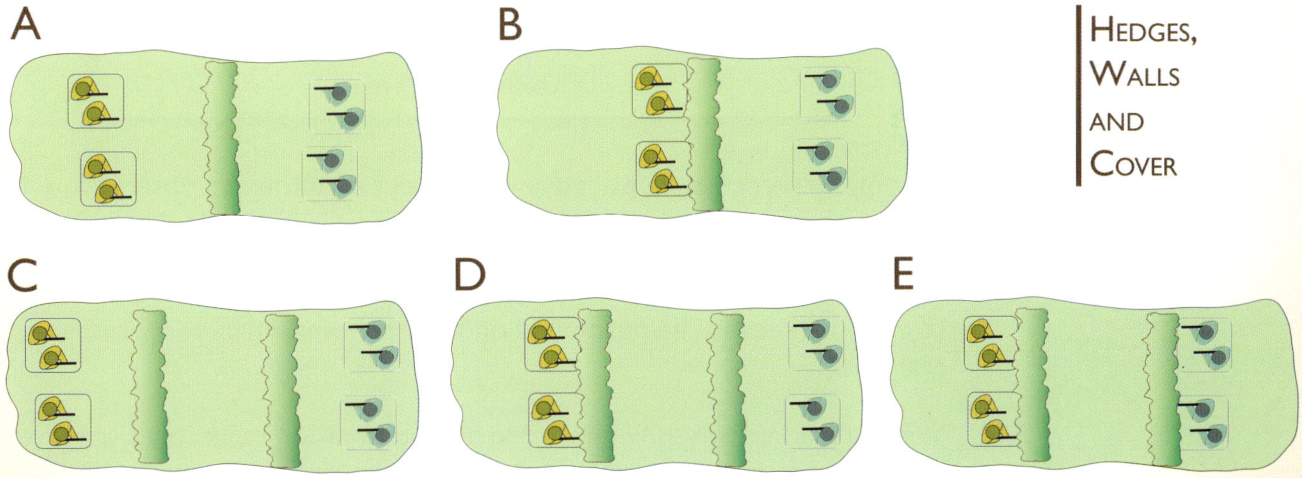

The above examples demonstrate when troops are, or are not, in cover:

- A The British (left) and the Germans are both in the open and can see each other.
- B The British (left) are in cover and the Germans are in the open.
- C The British and the Germans cannot see each other.
- D The British (left) are in cover and the Germans are in the open.
- E The British and the Germans are both in cover.

Note that the principles apply to walls as well as hedges.

Markers

2.13 If players wish to conceal elements from view that are beyond automatic observation range they may use markers to represent these models or figures on the table top. This lessens the chance of players' decisions being influenced by seeing units on the table that are supposed to be hidden (for example eight model figures of a German infantry company deep inside a wood can be physically seen by both players, but are considered to be hidden from enemy observation).

Markers can take many forms, but we would suggest pieces of 2" x 3" thin plywood decorated to match the terrain and topped with some foliage or netting to suggest camouflage.

Playing Tip

2.14 An individual marker may represent any one of the following:

- Up to a company (or equivalent) of infantry or cavalry, including HQs.
- A deployed gun or support weapon.
- A vehicle.
- A regimental or higher HQ.

2.15 The underside of the markers should be numbered (1-10 or 1-20 etc) if representing real units, or have 'dummy' written on them if not. The numbers should correspond to a list the player must make of any units being represented by markers. For example:

1 -	*Infantry company*
2 -	*MMG*
3 -	*Sherman 75mm*

Dummy markers can be placed anywhere out of automatic observation range to help deceive and confuse the enemy.

2.16 To represent the difficulty of spotting and identifying hidden units one dummy marker may be positioned on table for every marker representing a real unit.

2.17 When a marker comes within automatic observation range, or is observed after a successful chance observation test, it must be replaced by the model or models it represents, or be declared as a dummy. All dummy markers may be observed at the ranges for 'other units in cover' that have not fired in the previous move.

2.18 Dummy markers may not be moved, but other markers may be manoeuvred in exactly the same way as the units they represent.

2.19 'Real' markers may only be placed where there is sufficient room to deploy the model or models they represent. In the case of buildings and defensive positions place markers just in front of the feature.

Rule 2.8: all these OPs will be able to see over buildings and woods between them and the target they are trying to spot.

Smoke

3.1 **Smoke** prevents enemy forces from observing friendly forces either by 'masking' enemy positions or concealing friendly elements. The firing of smoke rounds takes place before other firing and movement, allowing troops to move behind the cover provided. Firing through smoke is not permitted.

3.2 The use of smoke rounds is limited to two rounds per game for guns (including all SP guns but not specialist towed anti-tank weapons, such as the British 6pdr) and mortars. Tanks guns that can fire HE may fire one smoke round per game.

3.3 A single smoke round creates a smoke cloud 3" wide by 1" deep. This may be represented on the tabletop by a cotton wool or kapok marker.

3.4 The aiming point for a smoke round must be within 2" and in front, or to the side, of an enemy element or position.

3.5 All weapons firing smoke may not fire HE (or AT) in the same turn.

3.6 The usual maximum ranges for weapons also apply when they fire smoke rounds.

Procedure for Firing Smoke

3.7 Nominate the gun / mortar that is firing smoke, identify the target element and where the smoke is being aimed, then throw a D6. If the result is a 5 or 6, a 3" wide smoke marker can be placed on the selected spot. Any other result is a miss. Repeat this procedure for all other weapons firing smoke this turn.

3.8 To create a multiple marker smoke screen, identify a target already hit by smoke and fire again, stating whether aiming to the left or the right of the existing marker(s). If they hit, place 'new' markers in contact with those already in place to form a straight line. Smoke screens may not curve around a target.

3.9 At the end of the enemy's next turn single smoke markers disappear and smoke screens of multiple markers drift 3", after which the marker nearest the direction of drift is removed. Determine the direction of drift with a D6 throw (1, 2, 3 left: 4, 5, 6 right), then move the smoke screen and remove the appropriate marker.

FIRING SMOKE

The German mortar has tried to fire two smoke rounds in front of the British MMG position, but D6 throws of 4 and 6 mean only one has hit. The 3" marker is wide enough to block observation of the German infantry and prevent both British reserved fire and fire in their next turn. Unless another weapon fires more smoke, this marker will be removed in the smoke stage of the next German turn.

4.1 In certain situations the player may want an AFV or infantry element (usually an anti-tank launcher team) to carry out one of the following heroic actions. These must always be tested for.

- When an AFV without infantry support (i.e. with no friendly troops within 6") wishes to advance to within 6" of observed enemy infantry.
- When an infantry element wishes to break cover and move in the open to attack an enemy AFV within 6".

4.2 A score of 6 is required on a D6 roll to carry out either of these actions, or 5 or 6 if the would-be 'heroic' element is elite. Elements which are part of units classed as poor for morale purposes cannot carry out these tasks.

4.3 If the AFV or infantry element fails this test it cannot carry out the action, or any further movement during this turn. However, it can carry out further tests in subsequent moves

4.4 Some specific (but rare) scenarios may allow infantry elements to carry out suicidal actions. For example, in a Pacific islands scenario a Japanese infantry element wearing demolition charges may wish to attack an American tank. In such cases both players must agree to suspend the normal heroic action tests.

An optimistic British PIAT team (regular morale) needs to leave the cover of the wood to attack the Panther tank. A D6 is thrown and the result is a 5. This means they have to stay in cover for this move.

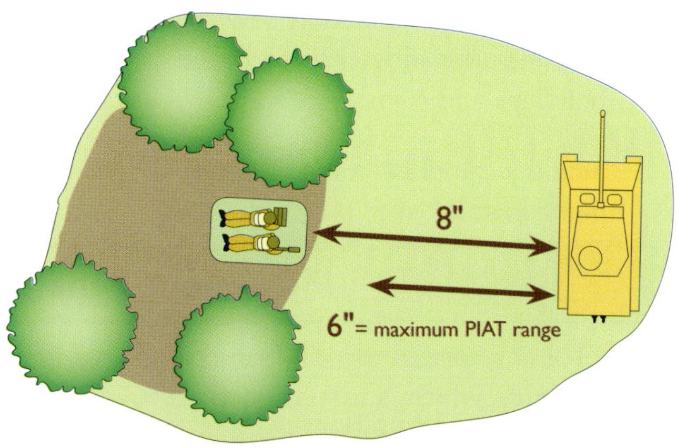

An Heroic Action

An American Bazooka team have passed their Heroic Action test and move to attack the unsuspecting Tiger II.

Table 6: Maximum Movement Distances

Category	Road*	Cross Country
Infantry	9"	6" (crawl 3")
Bicycles	15"	6"
Cavalry or horse mounted infantry	18"	12"
Horse drawn wagons and guns	15"	9"
Manhandled wheeled guns	6"	3"
Cars	30"	9"
Jeeps, Kubelwagens etc	30"	18"
Light trucks, armoured cars, half-tracks, Bren Carriers (or similar) and motorcycles	30"	15"
Other trucks and guns being towed by a suitable prime mover	24"	12"
Slow tracked vehicles	18"	9"
Tracked vehicles	24"	12"
Fast tracked vehicles	30"	15"

* To move at road speed a unit or vehicle must spend its entire turn on a road.
Note: amphibious vehicles may not move more than 4" a turn in water.

5.1 In built-up areas, such as towns and villages, maximum infantry speed is 6". This includes movement on roads, in squares, yards and gardens and inside buildings and ruins.

5.2 Infantry can crawl a maximum of 3" per turn. Crawling troops count as being in soft cover (see **rule 8.8**).

Movement Penalties

5.3 Crossing certain terrain features will result in a **Variable Movement Penalty**, which is tested for by throwing a D6 and subtracting the result from the element's **Maximum Movement Distance**. The terrain features fall into two groups:

- ***Linear***, such as walls and hedges.
- ***Continuous***, such as woods and hills.

5.4 Linear obstacles are diced for only once. If an element reaches a linear obstacle and still has some of its maximum movement distance left it may test to cross. If the result of subtracting the D6 throw from the remaining allowance is 1 or more the element may cross the obstacle and be moved forward the appropriate distance. If the result is 0, or negative, the element must wait to test in one of the movement stages of its side's next turn, when the penalty will be subtracted from its maximum movement distance.

5.5 It is not necessary to dice for individual figures crossing these features (such as runners with messages), but a test must be taken for each company-sized infantry / cavalry unit, support weapon, HQ, independent infantry element or vehicle.

5.6 Woods may be open or dense and both incur a variable movement penalty on units passing through them. Dense woods may only be crossed by infantry and cavalry. Vehicles and guns may be parked or emplaced just inside dense woods for cover if they or their bases remain in contact with the edge of the wooded area (see **rule 2.11**).

5.7 Unless gapped by an AFV or engineers, walls and hedges may not be crossed by cavalry, horse drawn transport or soft skin vehicles.

5.8 Once a dice is thrown the element must be moved Be aware that if an infantry element moving cross country (maximum 6" move) throws a 6 for trying to cross an obstacle no movement will take place (6" – 6 = 0)! If the result is 0 no further test is necessary to cross this obstacle in the following turn.

LINEAR OBSTACLES

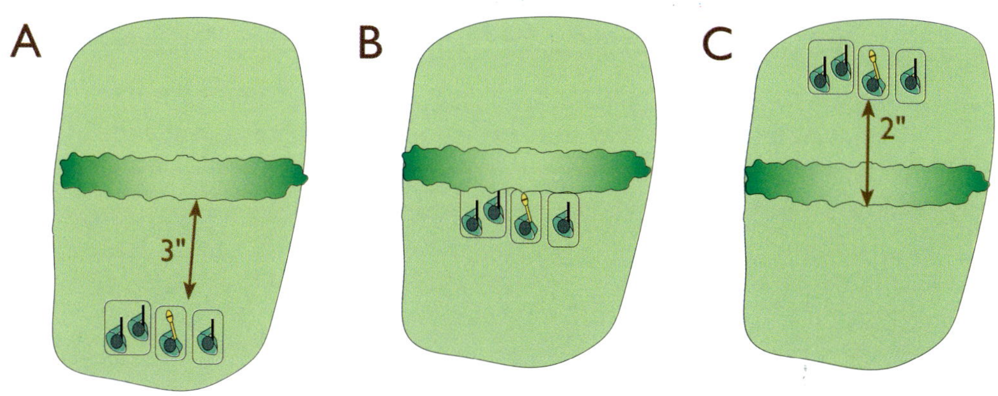

A A German infantry element moving cross country wishes to cross a hedge (a linear obstacle). They are 3" from the obstacle at the beginning of their turn.

B They use 3" of their movement allowance (6") to make contact with the hedge. A D6 is then thrown to discover the penalty to be deducted from their remaining movement. The throw is 1; meaning 1" is subtracted from the remaining 3", to give 2" of movement.

C The element crosses the obstacle (which is 1" wide in this case) and is placed 1" further on from the hedge.

5.9 Deduct 6" from normal vehicle maximum movement distances for infantry debussing (climbing out or off vehicles), or embussing (climbing in or on vehicles).

5.10 Deduct 3" from maximum movement distances for cavalry mounting or dismounting. Cyclists make no deductions for mounting or dismounting.

5.11 Infantry debussing and cavalry dismounting may move 3" and fire at short range.

Movement Penalties Continued

5.12 Cavalry may charge once per game, adding 6" to their normal cross country movement distance.

5.13 Deduct 6" from maximum movement distances for limbering or unlimbering towed guns and mortars. An additional 6" is deducted from a towing vehicle's movement if the towed gun is fired in the same turn it unlimbered. This is only available to guns on two wheeled carriages of 105mm or less. Other guns may not move, unlimber and fire in the same turn.

LIMBERING GUNS

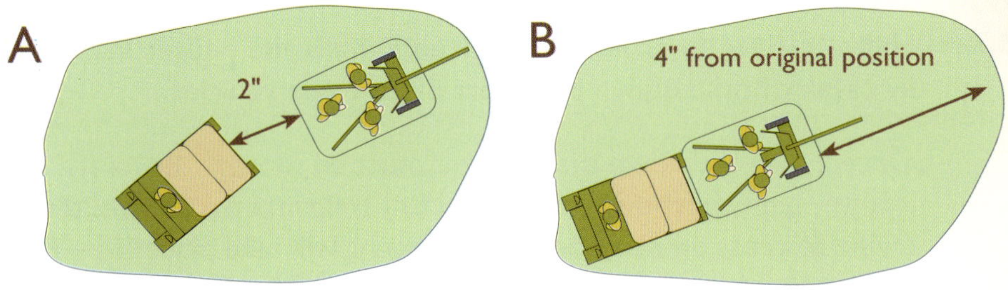

The Loyd Carrier prime mover has a cross country speed of 12".
A It uses 2" to reach, and 6" to limber the 6pdr AT gun.
B It may then move the weapon a further 4" from its original position.

5.14 There are no deductions for turning, but vehicles cannot reverse more than 6" in a turn (i.e. AFVs may not turn, move their full distance and turn again to face the enemy if they move more than 6").

Rule 5.1: all the American units moving through this town are restricted to a maximum move of 6".

6 AND 8. FIRING

	Short Range →													hard cover						soft cover						open						
	Medium Range →							hard cover						soft cover						open												
	Long Range →	hard cover						soft cover						open																		
	Dice throw =	1	2	3	4	5	6	1	2	3	4	5	6	1	2	3	4	5	6	1	2	3	4	5	6	1	2	3	4	5	6	
Points =	1																		1						1						1	1
	2																		1						1					1	1	2
	3												1						1						1				1	1	2	3
	4						1						1						1					1	2			1	1	1	2	4
	5						1						1					1	1				1	1	2			1	2	2	2	5
	6						1					1	1				1	1	2			1	1	1	2		1	1	2	2	3	6
	7						1				1	1	1				1	1	2			1	1	1	2		1	1	2	3	3	7
	8					1	1				1	1	2			1	1	2	2		1	1	1	2	3		1	2	2	3	4	8
	9					1	1			1	1	1	2			1	1	2	2		1	1	2	2	3		1	2	2	3	4	9
	10				1	1	2			1	1	1	2		1	1	2	2	3		1	2	2	3	3		1	2	3	4	4	10
	11				1	1	2		1	1	1	2	2		1	1	2	2	3		1	2	2	3	4		1	3	3	4	5	11
	12				1	1	2		1	1	1	2	3		1	1	2	2	3		1	2	3	4	4		2	3	3	4	5	12
	13				1	1	2		1	1	1	2	3		1	2	2	3	3		2	3	3	4	5	1	2	3	4	5	6	13
	14			1	1	2	2		1	1	2	3	3		1	2	2	3	4		2	3	4	5	5	1	2	3	4	5	6	14
	15			1	2	2	3		1	2	2	3	4		1	3	3	3	4		2	3	4	5	6	1	2	4	5	6	7	15
		Table 1						Table 2						Table 3						Table 4						Table 5						

Small Arms

Troops	Short Range	Medium / Long Range
Elite	Figs + 3	Figs + 2
Regular	Figs + 2	Figs + 1
Poor	Figs + 1	Figures

7. Close Assault

- Attackers: figures + D6
- Defenders: figures + D6
  D10 if in hard cover
  Elite + 1 / Poor - 1
  Difference = casualties
  (but maximum of 3 on loser and 1 on winner)

Machine Guns and Autocannon Values

Weapon	Minimum Crew	AT Gun Class	Small Arms Points Value		
			Ground Mounted	Vehicle Mounted	
				Stationary	Moving
Single LMG * / VMG **	1	None	6 *	6	4
Single MMG	2	None	8	8	6
Single HMG	2	6	10	10	6
Multiple LMG / VMG	2	None	8	8	6
Multiple MMG	2	None	12	12	8
Multiple HMG	2	6	15	15	10
20mm autonnon #	2	6	8	8	8
37mm autocannon #	2	5			
Multiple autocannon #	2	5	15	15	10

* LMGs only fire as an individual weapon if they are emplaced or vehicle mounted

** Co-axial turret-mounted VMGs cannot fire in the same turn as the main gun

\# If firing causes a crew casualty throw a further D6. 6 = weapon destroyed

Ranges

	Short	Medium	Long
Small Arms / most MGs	0 - 6"	6 - 14"	14 - 24"
HMGs / autocannon	0 - 6"	6 - 16"	16 - 30"
Anti-tank guns *	0 - 12"	12 - 30"	30 - 48"
Direct HE firing *	6 x 8" bands		
IDF	120" (72" for inf guns / MRLs)		
Light mortars	minimum 3"	maximum 18"	
Medium mortars	minimum 6"	maximum 48"	
Heavy mortars	minimum 6"	maximum 60"	

* Some high-powered weapons have different range bands and these are identified on the VG charts

To Hit With IDF / Mortars

Target →	Stationary	Moving
1st gun firing IDF or 1st mortar round	5 or 6	6
2nd / 3rd gun firing IDF or 2nd mortar round	4, 5 or 6	5 or 6

Achieving an Effective Hit From AT Firing

Gun Class →	1	2	3	4	5	6
Target ↓	D6 Throw Needed For an Effective Hit					
A class	4	5	6			
B class	3	4	5	6		
C class	2	3	4	5	6	
D class	2	2	3	4	5	6
E class	2	2	2	3	4	5
F class	2	2	2	2	3	4

Modifiers ↓	
-1	If the target or firer moved. Does not apply to AT rifles.
-1	If the target is partially concealed
-1	If the target is at long range.
+1	If the target is at short range.
+1	If the gun is firing at the rear of the target vehicle.

Damage From AT Hits

Range ↓	Damage to A to E class targets	F class
Short / medium	1 or 2 = light, 3 or 4 = heavy, 5 or 6 = destroyed	4, 5 or 6 = destroyed 1, 2 or 3 = no effect
Long	4 = light , 5 = heavy , 6 = destroyed	

Grenades

Target ↓	D6 throw and effect			
	1	2	3	4, 5 or 6
Figures *	1 casualty	2 casualties	3 casualties	No effect
Soft-skin vehicles **	No effect	Destroyed		
Armoured vehicles **	Light damage	Heavy damage	Destroyed	

* If casualty is a crew figure throw a further D6. 6 = weapon destroyed

** 5 or 6: To place grenade on stationary AFV / Throw HE grenade into an open-topped AFV / Hit open-topped AFV or soft-skin with rifle grenade (6 if moving).

RAPID FIRE! PLAYSHEET

1. MORALE TESTS

INFANTRY AND CAVALRY UNITS

TEST:
When the unit is at half strength / 20 casualties

D6 MODIFIERS:

-1	If CO is a casualty
-1	If unit is poor
-1	Unsupported
-1	Under artillery fire
-1	For each consecutive test
+1	If unit is elite
+2	50% of survivors are in hard cover
+2	All survivors are in defences

RESULTS:

4 or more	No effect
3	No offensive action
2	Pinned down
1 or less	Surrender / rout

INDEPENDENT AFVS AND ARMOURED UNITS

TEST:
When an AFV is heavily damaged / destroyed

D6 MODIFIERS:

-1	For each heavily damaged / destroyed AFV in unit
-1	Within 12" of enemy they can see in hard cover
-1	If unit is poor
+1	If unit is elite

RESULTS:

3 or more	No effect
2 or 1	No offensive action
0	Retire two moves
-1 or less	Surrender / rout

IDF AND ATTACKS BY AIRCRAFT

TEST:
When fired at by 80mm+ IDF / aircraft

D6 MODIFIERS:

-1	If unit is poor
+1	If unit is elite
+1	Infantry / gunners in hard cover

RESULTS:

4 or more	No effect
3 or more	Surpressed

INFANTRY AND CAVALRY IN THE OPEN

TEST:
When unsupported within 12" of enemy AFV

D6 MODIFIERS:

-1	If unit is poor
+1	If unit is elite

RESULTS:

4 or more	No effect
3 or more	Retire

2. OBSERVATION

TARGET	Fired last Move?	AUTOMATIC Normal	AUTOMATIC Recce	AUTOMATIC Artillery OP	CHANCE TEST (Soft: 5 or 6 / Hard: 6) Normal	CHANCE TEST Recce	CHANCE TEST Artillery OP
Infantry, support weapons and AT guns up to 75mm in the open	No	36"	42"	48"			
	Yes	48"	54"	60"			
Other targets in the open	No	48"	54"	60"			
	Yes	60"	66"	72"			
Infantry, support weapons and AT guns up to 75mm in cover	No	6"	12"	18"	6-12"	12-18"	18-24"
	Yes	36"	42"	48"			
Other targets in cover	No	12"	18"	24"	12-24"	18-30"	24-36"
	Yes	48"	54"	60"			
Infantry, support weapons in ruins	No	6"	12"	18"	6-12"	12-18"	18-24"
	Yes	48"	54"	60"			
Infantry, support weapons in intact buildings	No	No automatic observation			0-12"	0-18"	0-24"
	Yes	36"	42"	48"			
Dummy markers		12"	18"	24"	12-24"	18-30"	24-36"

3. SMOKE

5 or 6 needed to hit with smoke round and position marker.

4. HEROIC ACTION

For unsupported AFVs to approach observed infantry, or infantry to break cover to attack an AFV:
elite elements: 5 or 6 / regular elements: 6 / poor elements cannot do this.

5 AND 9. MOVEMENT

CATEGORY	ROAD	CROSS COUNTRY
Infantry	9"	6"
Bicycles	15"	6"
Cavalry or mounted infantry	18"	12"
Horse drawn wagons and guns	15"	9"
Manhandled wheeled guns	6"	3"
Cars	30"	9"
Light trucks, ACs, HTs, carriers and motorcycles	30"	18"
Other trucks and prime movers (when towing)	24"	12"
Slow tracked vehicles	18"	9"
Tracked vehicles	24"	12"
Fast tracked vehicles	30"	15"

MOVEMENT PENALTIES / RESTRICTIONS

- To move at road speed an element must spend its entire turn on a road
- Subtract 1 x D6 throw for each linear obstacle, or per turn, for continuous obstacles such as woods and hills.
- Infantry crawl a maximum of 3" per turn
- Subtract 6" from vehicle movement for embussing / debussing, limbering / unlimbering
- Subtract a further 6" if limbered / unlimbered gun fires
- Infantry movement through built up area is a maximum of 6"
- Debussing / dismounting troops may move 3" and fire at short range only

Movement and Other Rules

Above: a continuous obstacle (the hill) slows German infantry.
Left centre: glider troops crawl 3" after landing, and claim soft cover.
Right centre: this open wood is a continuous obstacle to the British recce vehicles.
Bottom: prime movers must be large enough to tow the desired weapon.

Reserved Fire

6.1 If an element reserves (holds) its fire by not moving or firing in its own turn it may interrupt the opposing player's turn after the first movement stage and fire at a suitable target

6.2 ***Reserved Fire*** may be used against any observed target: static or moving.

6.3 ***Reserved Fire*** may only be carried out by small arms firing groups, MGs and guns (but not mortars) able to engage in direct fire along a 'line of sight'.

Reserved Fire Procedure

6.4 Wait until the enemy have carried out all their movement (in the first movement stage only). Declare all elements that wish to carry out reserved fire and clearly identify both their individual targets and at what point during the enemy's move the shots were fired. For example, the moment when a vehicle crossed a gap between two buildings.

6.5 After firing, any vehicle that is destroyed or suffers heavy damage should be moved back to the point it had reached when the reserved fire fire took place.

6.6 The one exception to this order of play is when the target is a vehicle in a column or convoy moving along a road. In this case, the enemy's movement of units (during their first movement stage of their turn) should be interrupted immediately after the targeted model has been moved to a new position. Reserved fire is then carried out and if the target is destroyed, or immobilised by heavy damage, the vehicle is placed back at the position on the road where it was hit. All vehicles on the road behind it (or in front if reversing) may not then move past this point on the road until the obstruction is removed.

6.7 Assault guns and deployed anti-tank guns may only use reserved fire if the target has moved into their restricted arc of fire. See the diagram below for the definition of these restricted arcs.

Elements eligible to carry out reserved fire may be identified at the end of the turn in which they didn't move or fire by placing an 'ammo' box marker behind them. This is removed when they fire.

Playing Tip

Arcs of Fire

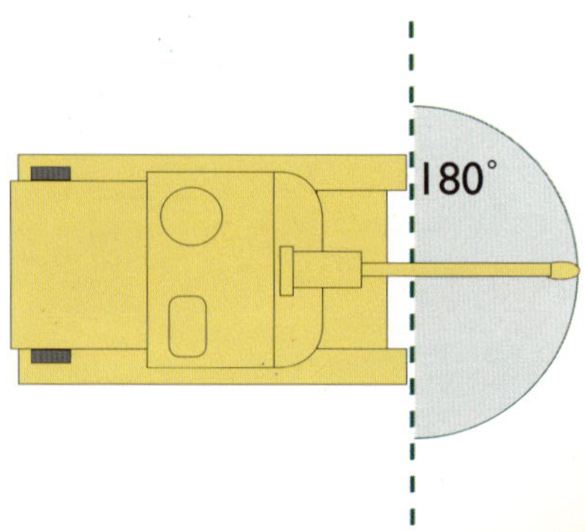

Deployed guns and vehicles with fixed weapons, such as assault guns and self propelled artillery, have fixed arcs of fire if they wish to engage targets without being considered as moving. The arc is 180° to the front of the vehicle or gun (measured from the front of the vehicle's hull or through the gun's axle). Deployed guns on cruciform mounts, such as the German 88mm flak gun and the British 2pdr AT gun, do not have a restricted arc of fire.

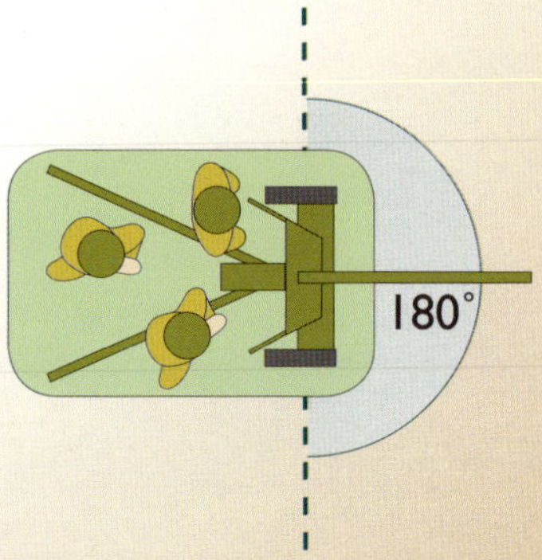

An Example of Reserved Fire

A

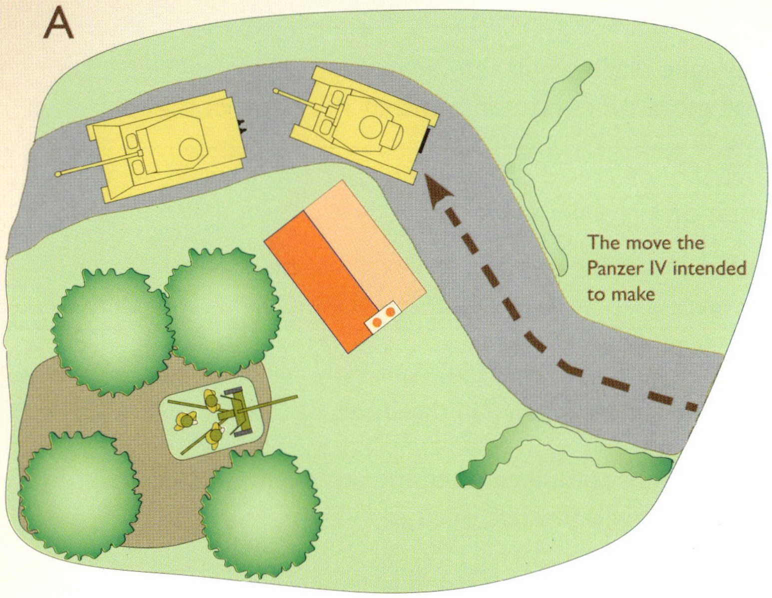

A

A column of three German vehicles (a Panther, Panzer IV and an Opel Blitz truck) moving along a road have to cross a gap between a hedge and a house that is overlooked by an unobserved British 6pdr AT gun concealed in a wood. The Panther is moved with no incident, but when the Panzer IV is moved the British player interrupts his opponent's turn and declares "I am going to have a reserved fire shot with my 6pdr at your Panzer IV as it moves across the gap".

B

The 6pdr fires and destroys the Panzer IV. The tank is placed on the road at the point where it was knocked out. The Opel Blitz cannot move any further along the road, but the Panther's movement is unaffected - at least until the morale throw!

B

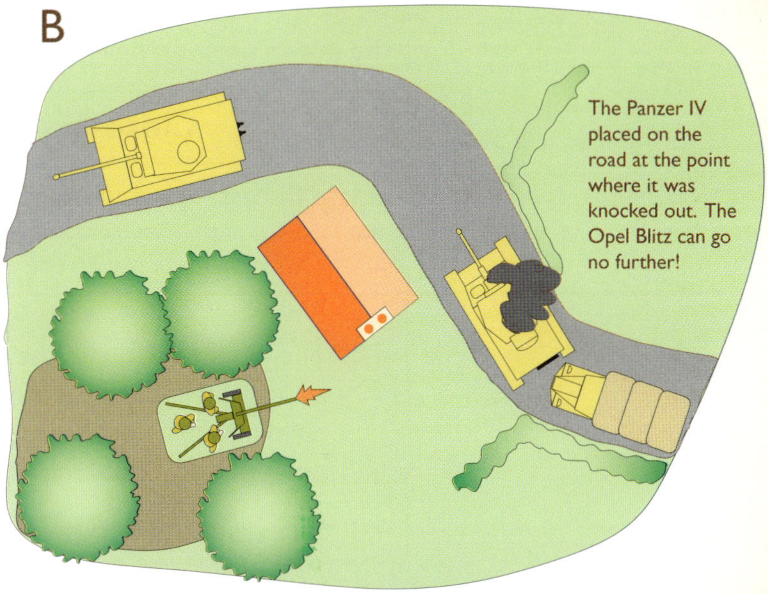

A British Firefly hits a Panther with reserved fire. causing heavy damage. Grey smoke can be used as an HD marker.

7.1 Close Assault may take place when the movement of an attacking infantry unit brings one or more of its elements within 1" of one of the following:

- Enemy infantry, or crews of deployed guns (or support weapons) in the open or in cover.
- A building, ruin or dug-in position occupied by enemy infantry, support weapons, or crews of deployed guns.

7.2 Cavalry can also carry out a close assault, but only against infantry, crews of deployed weapons and other cavalry in the open.

7.3 The close assault test must be taken after the attackers have completed their move and any reserved fire has been resolved.

7.4 Attacking figure bases may not be included when counting those involved in close assault if they are over 1" away from either: one of the enemy figure bases; or the edge of the defended building, ruin, or cover (such as a wall), or the dug-in position being attacked.

Close Assault Procedure

7.5 Calculate the assaulting and defending scores:

- The attacker counts the assaulting figures and adds this number to a D6 throw.
- The defender counts the number of figures receiving the assault, adding it to a D10 throw if they are occupying hard cover, buildings, ruins or dug in positions. If the defenders are in the open or soft cover then throw a D6.
- Elite units add 1 to these throws. Poor troops subtract 1.

7.6 The highest score decides who has won the combat and the difference between the scores is the number of casualties removed from the losing unit, up to a maximum of three. The winning unit loses one figure.

7.7 The loser must immediately withdraw a full move away from the winner towards a friendly unit, or to the nearest cover not occupied by the enemy. The winner may immediately occupy the defender's positions, including any buildings or ruins, but may not move more than 6" from their original assault position.

7.8 If occupation of a vacated position brings assaulting troops into close assault range of another enemy unit (such as a second company, or a support weapon not involved in the close assault) no action can be taken until the defender's turn. The defenders may then choose to fight or withdraw a full move. No additional figures may be added to a defending unit if it decides to fight.

7.9 If close assault results in a draw, the attacker must withdraw a full move towards a friendly unit, or to the nearest cover not occupied by the enemy.

Cavalry Close Assaults

7.10 Mounted cavalry rarely got close to enemy troops in WWII, but it did happen occasionally. Cavalry may only close assault other cavalry or infantry in the open (but not those occupying any sort of defensive position).

7.11 For cavalry attacking cavalry, each side adds the number of figures to a D6 roll. For cavalry attacking infantry the cavalry adds the number of figures to a D10 roll and the infantry add the number of figures to a D6 roll.

Examples of close assault against a deployed gun in a ruin and (below) infantry in the open.

Close Assaults on Buildings

7.12 An individual model building or ruin (which may also include a pavement, a garden or courtyard) represents an area that can be occupied and defended by infantry and, if defended, assaulted by infantry. Unless specified before the game begins, all single buildings of small or moderate size (such as individual houses, shops or barns) may be occupied by up to eleven figures. The capacities of larger buildings (such as churches, or interconnected farm buildings) should be agreed before the game.

7.13 Very large buildings (such as factories) and connected buildings (such as terraced houses) may be divided into a number of separate areas for close assault. Decide how many and where their boundaries are before the game begins.

7.14 Infantry and support weapons occupying a building do not need to be physically in contact with all the outside walls, but are assumed to be capable of defending against an attack from any direction. If the defenders did not move into the building or fire from it in their last turn they may use reserved fire against any attack.

EXAMPLE OF CLOSE ASSAULT ON A BUILDING

A The German assault objective is a house defended by seven regular British infantrymen. The defenders did not move or fire in their previous turn and have reserved fire capability.

B **Stage 5** of the German turn. Nine regular German infantrymen move up to make contact with two sides of the defended building.

C **Stage 6** of the German turn. The British carry out their reserved fire and kill one of their attackers.

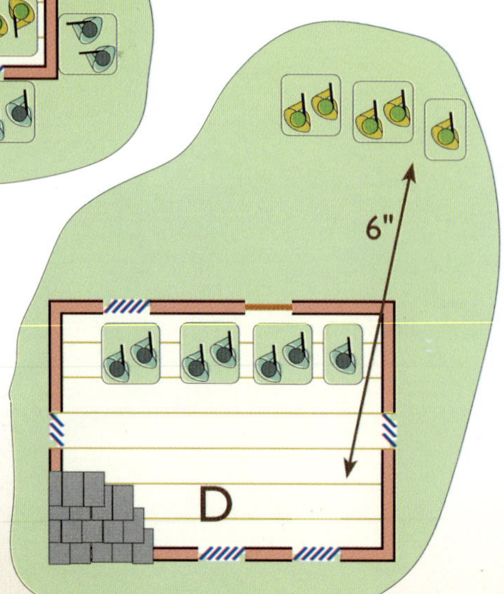

D **Stage 7** of the German turn - resolving the close assault. The seven defenders throw 4 on a D10 (total = 11) and the eight surviving attackers throw 6 on a D6 (total 14). As the attackers won by a margin of three they inflict three casualties on the defenders and suffer one casualty themselves. The British losers withdraw a full move and the German victors occupy the building.

7.15 Buildings, ruins and dug-in positions may only be assaulted on one or two sides at a time or, if the base has an irregular shape, along no more than 50% of the perimeter.

Rule 7.13: the terrace of buildings above may be divided into separate areas for close assault. In the village below the larger building, with agreement, may hold more than eleven figures.

Firing Elements

8.1 A firing element may be one of the following:

- A small arms firing group.
- A deployed gun or rocket launcher (including those in, or on vehicles).
- A support weapon, such as an MG, mortar or anti-tank rocket launcher.

Targets

8.2 Before any firing takes place players must identify all elements that wish to fire at a particular target. Once firing has started no other elements may be added to those originally identified. If a target has been destroyed before all fire directed at it has been completed the remaining firing elements cannot switch to other targets. A target may be any of the following:

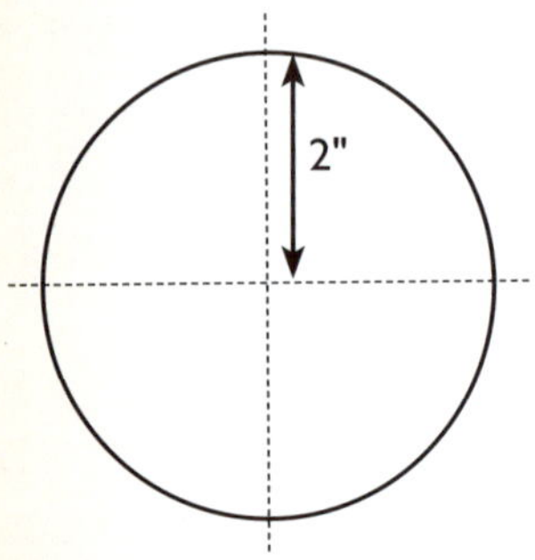

- Figures and weapons in a 4" diameter **Risk Zone** centred on a single identified element such as a figure on foot, horse or bicycle, deployed gun, multiple rocket launcher, or support weapon.
- A vehicle.
- A building*, ruin or bridge, including pontoon bridges under construction.
- An emplacement, bunker, or section of trench up to 4" long.
- Special, mutually agreed targets such as aircraft on the ground, boats, ammunition dumps etc.

* Only individual buildings or ruins where troops have been observed (or observed entering or firing from) may be targeted. Bridges are allowed as targets because of their individual tactical importance.

> To help measure HE risk zones, cut a 4" diameter circle of clear plastic and mark the centre with a red dot to help position it over the centre of targets.
>
> Playing Tip

Measuring Ranges from the Firing Element to the Target

8.3 No measuring of ranges is permitted before making a decision to fire.

8.4 Ranges (whether for observation or firing) are measured from the nearest edge of the firing element's base (or the closest base if there are more than one, for example in a small arms firing group) to:

- The centre point of the risk zone.
- The nearest point of the vehicle (or base* if the model has one).
- The nearest point of the building or other target.

8.5 A miss has no effect. On our scaled-down battlefield shots which miss their intended target would stand too high a chance of hitting something else.

Examples of Measuring Ranges and Applying the Risk Zone

The red arrows show the points from which ranges measurements should be taken.

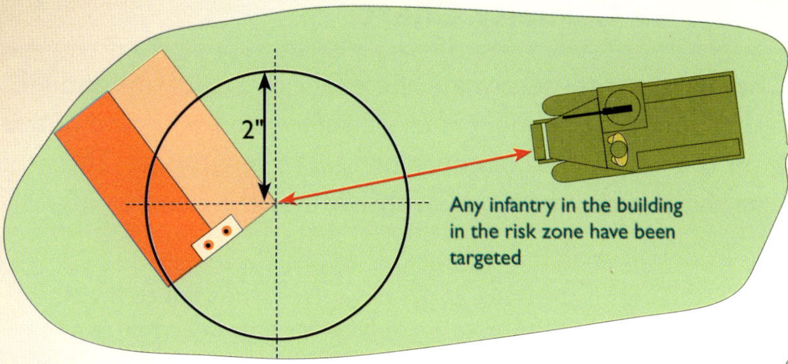

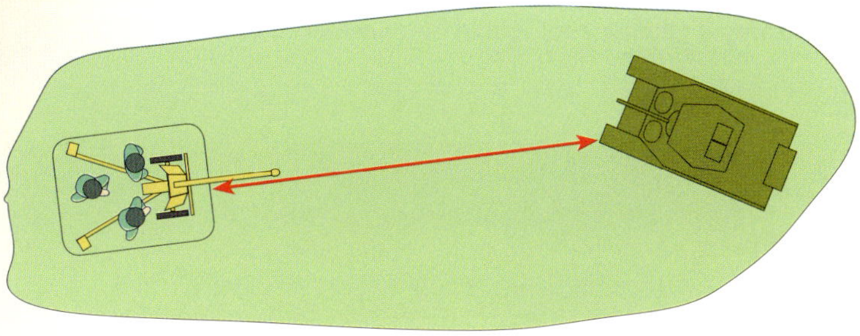

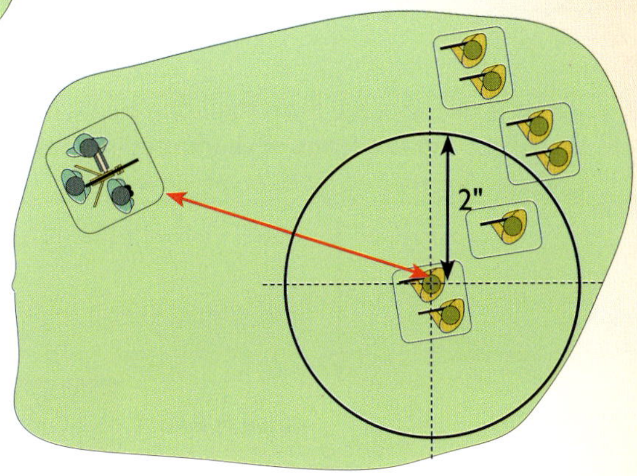

The circle shows the 4" diameter risk zone centred on an individual figure and demonstrates that the MMG has been able to target three figures.

Minimum and Maximum Crew Levels

8.6 Towed guns, MGs, mortars and other guns require a minimum number of figures. A maximum number of crew figures is also necessary to ensure that weapons are not 'over manned'.

Weapon	Minimum Crew	Maximum Crew
Ground-mounted MMGs and HMGs	2	3
Light mortar / AT launcher / man-pack flamethrowers	1	2
Medium mortar	2	3
Heavy mortar	2	4
Gun up to 79mm	2	4
Gun of 80mm and over	2	5
Multiple rocket launcher	2	3

- Any anti-tank gun or support weapon may be re-crewed by any 'friendly' figure. Unless the game scenario specifically allows it, no enemy equipment can be re-crewed.
- Artillery unit guns and multiple rocket launchers can only fire IDF if at least one of the two minimum crew is a specialist gunner.

The Casualty Chart

8.7 The **Casualty Chart** is used to determine the effect of small arms and HE fire. The points value of a firing element is cross-referenced with one of five tables, depending on the range, type of cover, or calibre of weapon. The chart then converts the score of a single D6 throw into the number of enemy casualties.

A range is the distance from the firing element to the target and the spans of short, medium and long are determined by the type of weapon.

Cover is the status of the target (see **rules 8.8 to 8.12**). If elements in a risk zone are in different states of cover (e.g. some are in soft cover and others are in hard cover) then assume that all elements in the risk zone are in the lower category (e.g. for the example all elements would be considered to be in soft cover).

Casualty Chart

	Table 1						Table 2						Table 3						Table 4						Table 5						
Short Range													hard cover						soft cover						open						
Medium Range							hard cover						soft cover						open												
Long Range	hard cover						soft cover						open																		
Dice throw =	1	2	3	4	5	6	1	2	3	4	5	6	1	2	3	4	5	6	1	2	3	4	5	6	1	2	3	4	5	6	
Points = 1																		1						1						1	1
2																		1						1					1	1	2
3												1						1						1				1	1	2	3
4						1						1						1					1	2			1	1	1	2	4
5						1						1					1	1				1	1	2			1	2	2	2	5
6						1					1	1				1	1	2			1	1	1	2		1	1	2	2	3	6
7						1				1	1	1				1	1	2			1	1	1	2		1	1	2	3	3	7
8					1	1				1	1	2			1	1	2	2		1	1	1	2	3		1	2	2	3	4	8
9					1	1			1	1	1	2			1	1	2	2		1	1	2	2	3		1	2	2	3	4	9
10				1	1	2			1	1	1	2		1	1	2	2	3		1	2	2	3	3		1	2	3	4	4	10
11				1	1	2		1	1	1	2	2		1	1	2	2	3		1	2	2	3	4		1	3	3	4	5	11
12				1	1	2		1	1	1	2	3		1	1	2	2	3		1	2	3	4	4		2	3	3	4	5	12
13				1	1	2		1	1	1	2	3		1	2	2	3	3		2	3	3	4	5	1	2	3	4	5	6	13
14			1	1	2	2		1	1	2	3	3		1	2	2	3	4		2	3	4	5	5	1	2	3	4	5	6	14
15			1	2	2	3		1	2	2	3	4		1	3	3	3	4		2	3	4	5	6	1	2	4	5	6	7	15

The points column is the factor determined by the size, quality and range of the firing group (e.g. six regular troops firing at a target 13" away would be 7 points).

The highlighted rows are those with are also used for assessing the damage from HE and mortar hits (e.g. a 3" mortar round uses 8 points on table 4). See **rules 8.61 and 8.74.**

A Luftwaffe ground crew small arms firing group.

Soft and Hard Cover

8.8 Figure elements and support weapons behind (and in contact with) a hedge, or in woods, or crawling are considered to be in soft cover. Guns, mortars and small arms firing from inside woods are in soft cover, but can only fire if they (or their bases) are in contact with the edge of the wooded area (see also **rules 2.11 and 5.6**).

8.9 Crew figures and passengers of soft-skin vehicles are in soft cover.

8.10 Infantry occupying buildings, ruins, trenches or sheltering behind (and in contact with) walls are in hard cover.

8.11 Crew figures and passengers inside open-topped AFVs are in hard cover. Infantry and other figures being carried on top of AFVs are in the open.

8.12 Gunners operating a weapon with an armoured shield are considered to be in hard cover from small arms and MGs firing at them from anywhere within an angle of 180° to their front, but they are classed as in the open if targeted by any form of HE fire.

Examples of soft cover (above) and hard cover (right).

Testing for Battalion CO or Artillery Observer Casualties

8.13 If a target that includes a Battalion CO or an Artillery Observer Team (OP) loses more than one casualty to any sort of fire, throw another D6 (two throws if both are present). A 6 means that the CO (or one of the OP team) is a casualty. This test is carried out each time an element which includes any of these specialist figures suffers losses.

A British small arms firing group fires at a radio operator in a German HQ group causing one casualty. As the German CO figure is within the 4" diameter risk zone he is potentially this casualty. A D6 is thrown and the result is a 5. Therefore the CO is OK and one of the other figures has been hit and removed.

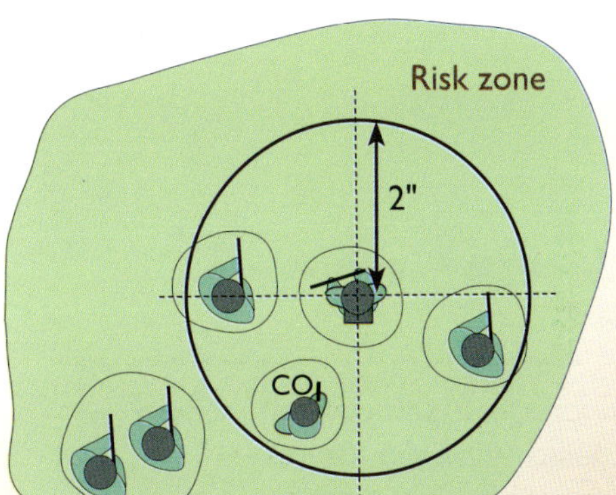

Assessing if a CO is a Casualty

Small Arms Firing Groups

8.14 A **Small Arms Firing Group** is a rifle company or similar unit (such as a troop of commandos), an HQ group or members of a support company who are not manning an infantry support weapon.

8.15 Each firing group can be made up of figures armed with rifles, light machine guns, sub-machine guns and pistols. In the basic rules it does not matter what small arms individual model soldiers are equipped with (see **advanced rule 13.1**), only the number of figures counts.

8.16 A battalion or similar unit may deploy a maximum of five small arms firing groups in any one turn, none of which may not be larger than twelve figures.

8.17 To calculate the firing points value, count the number of figures that have a clear line of sight to the target (see **rules 2.7 to 2.13**) and add the appropriate factors for troop type and range from **Table 7** below.

8.18 To be counted as being part of the same small arms firing group, each figure base (single or multiple) must be within 1" of at least one other figure base belonging to the same group.

8.19 There is no need to test for a hit. Using the casualty chart, calculate the points value of the **s**mall arms firing group, consult the appropriate table and throw a single D6.

8.20 Casualties caused by small arms fire have no effect on deployed guns and support weapons, only their crews.

TABLE 7: SMALL ARMS FIRING GROUPS

Troop Type	Short Range 0 – 6"	Medium Range 6" – 12"	Long Range 12" – 18"
Elite	Figures + 3	Figures + 2	Figures + 2
Regular	Figures + 2	Figures + 1	Figures + 1
Poor	Figures + 1	Figures	Figures

SMALL ARMS POINTS VALUES

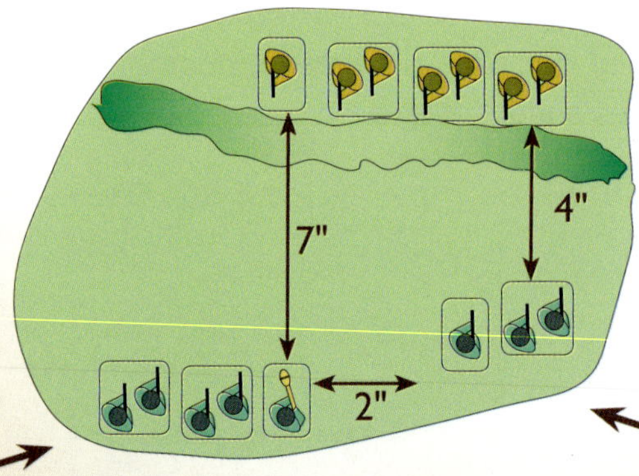

Two groups of regular German infantry (five figures and three figures) from a rifle company are separated by more than 1" and therefore must form two small arms firing groups. As the figure with a Panzerfaust is not firing it this turn he may be added to the total of firing figures in his group.

This group is firing at a target in soft cover at medium range. It has a total points value of 5 (figures) + 1 (addition for regular troops firing at medium range) = 6.

This group is firing at a target in soft cover at short range. It has a total points value of 3 (figures) + 2 (addition for regular troops firing at short range) = 5.

Grenades

8.21 As an alternative to using small arms, figures in a small arms firing group may throw grenades: either AT or HE, but not both during the same turn.

8.22 One grenade may be thrown per four figures. Groups of less than four figures may not use grenades. The four figures that are involved in throwing a grenade may not form part of a firing group this turn.

8.23 The maximum grenade range is 3". AT grenades are only effective if placed on a stationary armoured vehicle by a figure in contact, or if successfully thrown into a moving open-topped armoured vehicle for a roll of a 5 or 6 on a D6 throw. See **Table 8** below for effects of grenades.

8.24 Rifle grenade launchers are classed as support weapons with a crew of one. They have a minimum range of 3" and a maximum range of 12". A D6 roll of 6 is required to hit a moving target and 5 or 6 to hit a stationary target. Rifle grenades have no effect on fully armoured (as opposed to open-topped) vehicles.

8.25 When throwing grenades into buildings at least one figure of the attacking unit must be within 1" of one of its walls. Ruins are treated as all other targets.

Table 8: Grenades

	D6 Throw and Effect			
Target	1	2	3	4, 5 or 6
Figures*	1 casualty	2 casualties	3 casualties	No effect
Soft skin vehicles	No effect	Destroyed		
Armoured vehicles**	Light damage	Heavy damage	Destroyed	

* If crews of deployed guns or support weapons lose casualties, their gun, mortar or AT launcher may be at risk. Throw a further D6. 6 = gun or support weapon destroyed.

** AT grenades placed on a stationary, enclosed AFV, or HE grenades successfully thrown into an open-topped AFV.

Grenades in Action

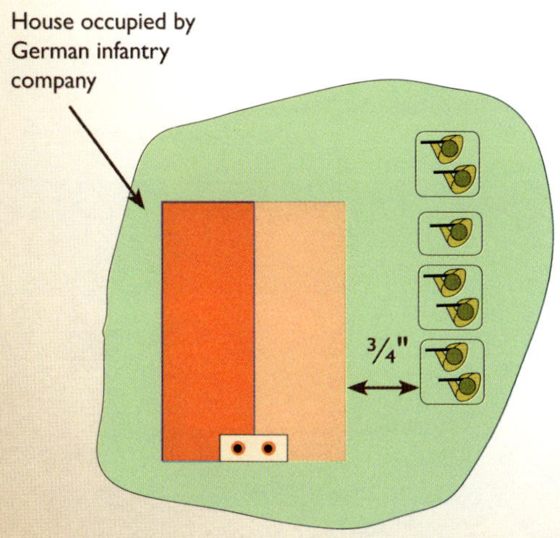

Seven British infantry advance to within 1" of a building occupied by an observed German company of eight figures (one of which has a Panzerfaust.) The British decide to throw in a grenade.

The D6 throw of 2 means the German company loses two figures and, due to the casualties, the Panzerfaust is at risk. A further D6 throw of 6 means it has been destroyed.

Machine Guns and Autocannon

8.26 Machine guns (MGs when talked about generally) fall into several categories. For the purposes of the rules these are grouped under four headings.

LMG	A light machine gun such as the British Bren and Lewis gun, the US Browning Automatic Rifle (BAR) or the German MG 34 / 42.
MMG	A medium machine gun such as the British Vickers, the US tripod-mounted 30cal Browning and the German MG 34 / 42 on a sustained fire mounting.
HMG	A heavy machine gun such as the US 50cal Browning, the British Besa 15mm or the German 15mm used in the 'Drilling' triple AA mount.
VMG	A vehicle machine gun on either a pintle or co-axial mounting (most commonly an MMG).

Many of these weapons were also used in multiple (principally AA) mounts of up to four guns.

8.27 Other, heavier automatic weapons fall into the autocannon category, such as the German 20mm and 37mm AA guns and British Polsten 20mm. These weapons also fire explosive rounds and can destroy deployed guns and support weapons (see **Table 9**). These weapons were also used in multiple mounts.

8.28 Vehicle mounted machine guns (VMGs) have the same ranges as small arms firing groups. To be treated as stationary a vehicle mounting an MG or other automatic weapon must not move during the turn in which it fires.

Below and bottom right: examples of LMGs counting as individual weapons.
Right: this Drilling AA gun is a multiple HMG.

Table 9: Machine Guns and Autocannon

<table>
<tr><th rowspan="3">Weapon</th><th rowspan="3">Anti-Tank Gun Class</th><th colspan="3">Ranges</th><th colspan="3">Points Value for Firing Effect</th></tr>
<tr><th rowspan="2">Short Range</th><th rowspan="2">Medium Range</th><th rowspan="2">Long Range</th><th rowspan="2">Ground Mounted</th><th colspan="2">Vehicle Mounted</th></tr>
<tr><th>Stationary</th><th>Moving</th></tr>
<tr><td>Single LMG* / VMG**</td><td>None</td><td rowspan="7">0 - 6"</td><td rowspan="7">6" - 16"</td><td rowspan="7">16" - 30"</td><td>6*</td><td>6</td><td>4</td></tr>
<tr><td>Single MMG</td><td>None</td><td>8</td><td>8</td><td>6</td></tr>
<tr><td>Single HMG***</td><td>6</td><td>10</td><td>10</td><td>6</td></tr>
<tr><td>Multiple LMG / VMG</td><td>None</td><td>8</td><td>8</td><td>6</td></tr>
<tr><td>Multiple MMG</td><td>None</td><td>12</td><td>12</td><td>8</td></tr>
<tr><td>Multiple HMG</td><td>6</td><td>15</td><td>15</td><td>10</td></tr>
<tr><td>20mm autocannon#</td><td>6</td><td rowspan="2">8</td><td rowspan="2">8</td><td rowspan="2">8</td></tr>
<tr><td>37mm autocannon#</td><td>5</td><td rowspan="2">0 - 12"</td><td rowspan="2">12" - 30"</td><td rowspan="2">30" - 48"</td></tr>
<tr><td>Multiple autocannon#</td><td>5</td><td>15</td><td>15</td><td>10</td></tr>
</table>

* LMGs are only treated as an individual weapon if they are in emplaced positions (such as bunkers and tobruk pits) or vehicle mounted. If this is not the case then they form part of a small arms firing group. They require a minimum crew of one.

** An AFV's co-axial turret-mounted VMG cannot be used in the same turn as the main gun is fired.

*** An anti-aircraft HMG mounted on an AFV may only againgst aircraft (see **advanced rule 15.26**).

\# If figures are crewing a deployed gun or support weapon targeted by an autocannon and a casualty is inficted throw a further D6 for the weapon. 6 = destroyed.

The Effect of Small Arms and MG Fire Against Vehicles

8.29 Soft-skin vehicles are treated as targets in soft cover (see the casualty chart) when fired at by small arms and machine guns. If a vehicle sustains two casualties 'worth' of damage from firing in one turn the vehicle is destroyed and the driver lost. Any additional casualties inflicted during the same turn are removed from the vehicle's passengers.

8.30 If a soft-skin vehicle only sustains one casualty 'worth' of damage from HE fire or small arms fire during a turn there is no effect on the vehicle. This casualty is not carried forward to the next turn.

8.31 A weapon mounted on, or towed by a soft-skin vehicle destroyed by small arms fire only is undamaged and may still be fired.

8.32 Small arms fire cannot destroy open-topped AFVs and APCs but can cause casualties amongst crew and passengers. These vehicles provide hard cover (see **rule 8.11** and the casualty chart) and AFVs have a nominal crew of three. If all three crew members of an open-topped AFV are lost it cannot take any further part in the game (and counts as destroyed for morale tests).

Guns

8.33 Guns fall into several different categories, including: anti-tank (AT); anti-aircraft (AA); dual or multi-purpose (such as the German '88'), and field, medium and heavy artillery. Guns may be towed, or mounted in self-propelled guns, assault guns and tank turrets. For the purpose of the rules what is important is:

- Anti-tank (AT) capability.
- High explosive (HE) capability.
- Ability to fire HE direct and/or indirect (IDF).
- Anti-aircraft (AA) capability.

8.34 Relevant AT and HE capabilities are set out in vehicle and gun tables (**VG Charts**).

8.35 IDF is only an option for guns that were traditionally used for that purpose, such as the British 25pdr field gun or German 105mm howitzer. Assault guns and tanks are not normally be used for IDF. Some weapons were only used in the IDF role and the VG charts indicate where this was the case.

8.36 AA fire is only an option for weapons that were designed for that purpose, such as the Bofors 40mm AA gun (see **advanced rules 15.26**).

Vehicle and Gun Classifications Tables (VG Charts)

8.37 VG Charts include towed, self propelled and tank guns.

8.38 The following diagram illustrates how a VG chart is constructed. Pages 6 to 21 include VG charts for Allied and German units in mid 1944.

VG CHART BREAKDOWN

A description of the vehicle.

The maximum vehicle movement shown as road / cross-country distances.

The class of the vehicle when determining an effective hit from AT firing.

The AT, HE, and MG values of the weapons mounted in the main body of the vehicle and its turret. Vehicles may not fire **Hull (and Other) Weapons** 'line-of-sight' from hull down positions (see **rule 8.44**).

Vehicle (Main Weapon)	Speed	Armour	Hull (and Other) Weapons	Turret Weapons
Sherman tank (75mm)	24" / 12"	C	VMG	AT class 3 / HE 10 points table 4 (or VMG)
M7 self-propelled gun (105mm)	24" / 12"	E	AT class 3 / HE 12 points table 4	
Gun				
57mm anti-tank gun	AT class 3 / HE 6 points table 4			

The AT and HE values of a gun. If no HE value is listed then the gun cannot fire HE ammunition.

Vehicles may only fire one **Hull (and Other) Weapon** and one **Turret Weapon** each turn. Any alternative to firing the main weapon is shown in brackets.

Examples of Artillery Batteries / Battalions

Top: a deployed British 25pdr battery.
Centre left: German 105mm battalion.
Centre right: British SP 25pdr battery.
Bottom. American SP 105mm battalion.

Anti-Tank Firing

8.39 All anti-tank weapons other than hand held launchers like Panzerfausts and PIATs have a short range, a medium range and a long range.

Short - 0 - 12" / Medium - 12 - 30" / Long - 30 - 48"
Note a VG chart may indicate different ranges for some high / low powered weapons such as the British 17pdr.

8.40 Anti-tank guns and other weapons with anti-tank capability, such as field guns, are classed from 1 to 6. 1 is the most powerful (such as an 88 mm PAK 43) and 6 is the least effective (such as an anti-tank rifle).

8.41 Vehicles are classified from A to F, with A being the best armoured and E the poorest. F vehicles are soft-skin vehicles with no armour. To damage or destroy a vehicle it is necessary to achieve an effective hit.

Achieving Effective Anti-Tank Firing Hits on Vehicles

8. 42 **Table 10** shows the minimum D6 throw needed to achieve an effective AT hit and also lists modifiers to this score. An effective hit means that the gun has hit the vehicle and has the potential to damage it.

TABLE 10: ACHIEVING AN EFFECTIVE HIT FROM ANTI-TANK FIRING

Gun Class →	1	2	3	4	5	6
Target	**D6 Throw Needed For an Effective Hit**					
A class	4	5	6			
B class	3	4	5	6		
C class	2	3	4	5	6	
D class	2	2	3	4	5	6
E class	2	2	2	3	4	5
F class	2	2	2	2	3	4

Modifiers (add or subtract from the dice throw)	
-1	If the target or firer moved in the previous turn, or if the firer has moved, or is about to move in the current turn. This penalty does not apply to anti-tank rifles.
-1	If the target is partially concealed (see **rule 8.44**).
-1	If the target is at long range.
+1	If the target is at short range.
+1	If the gun is firing at the rear of the target vehicle (see **rule 8.45**).

8.43 A hull-down tank is partially concealed and may fire its turret gun. To claim this type of cover most of a tank's hull must be hidden behind a wall, earth bank or hill / ridge with only the turret in view. In the case of a hill, the tank must be within 3" of the highest point and on the opposite side to the enemy.

8.44 Vehicles are classed as partially concealed if more than half of the vehicle is hidden from the AFV or gun firing at them.

Partially Concealed Vehicles

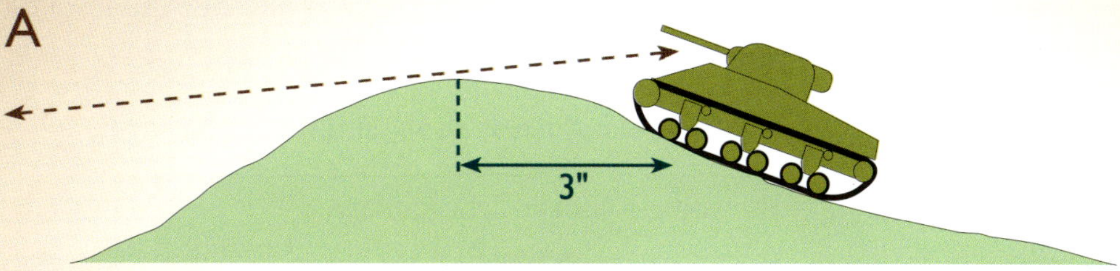

Situations classified as partially concealed:

- A 'Hull down', in which a tank uses a ridge or fold in the ground to make only its turret visible to the enemy. This firing position is not available to vehicles with hull-mounted guns such as a StuG III assault gun or the US Lee / Grant tank.
- B Vehicles behind and in contact with hedges and walls.
- C Vehicles at the edges of woods that have have been observed by the enemy.
- D When more than 50% of the vehicle has been obscured by an object such as a building or another vehicle.

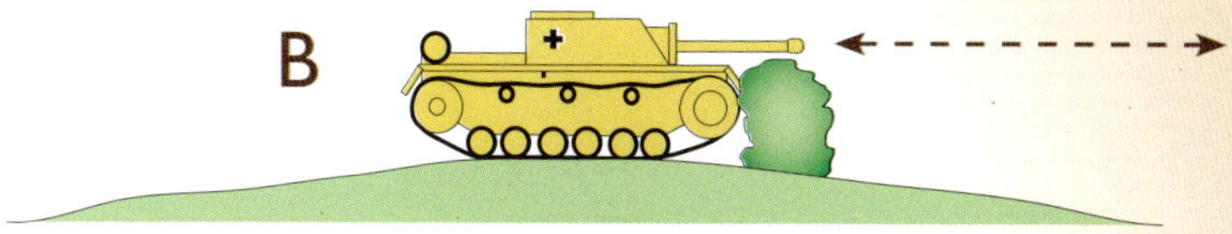

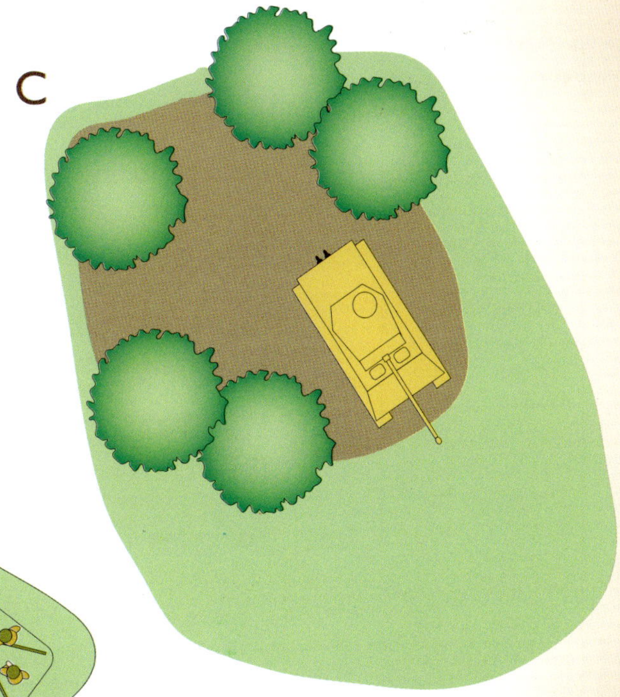

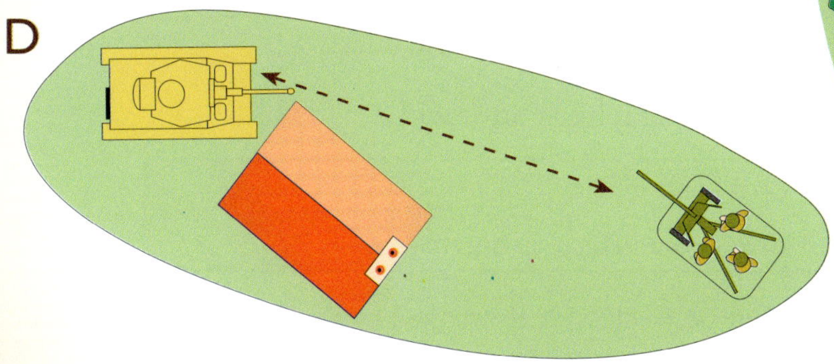

8.45 To fire at the rear of an AFV, a gun must be behind a line drawn laterally through the rear of the AFV's hull (see below) and, if the gun is on a two wheeled carriage, the target must be in front of a line drawn through the gun's axle (see **rule 6.7** and below).

Defining a Rear Shot

A shot at the rear of a vehicle is one where the firer is behind a line drawn laterally though the rear of the vehicle's hull. In the diagram the 75mm AT gun is behind this line but the Panzerschrecke at B is in front and does not have a rear shot.

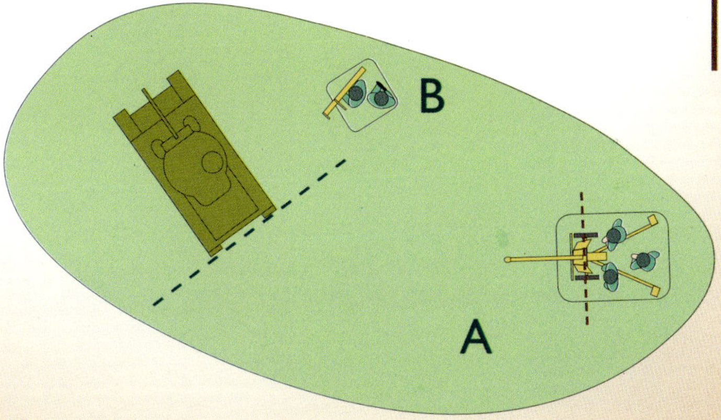

Assessing Damage To Vehicles From an Effective AT Hit

8.46 Once an effective hit is achieved throw another D6 to find out the level of damage to the target vehicle.

8.47 For armoured vehicles (A to E class) hit at short and medium range:

> 1 or 2 = the vehicle suffers light damage.
>
> 3 or 4 = the vehicle suffers heavy damage.
>
> 5 or 6 = the vehicle is destroyed.

8.48 At long range the chances of damage being caused by an effective hit are halved:

> 1, 2 or 3 = no effect.
>
> 4 = light damage, 5 = heavy damage, 6 = destroyed.

Anti-Tank Gun Portees

Portees are (usually British) soft-skin trucks carrying anti-tank guns. Portees move at truck speeds. The gun can be fired from the portee, but not on the move. The gun is lost if the truck is destroyed. The AT gun may also be dismounted for firing, for a deduction of 6".

8.49 Damage to an armoured vehicle is cumulative, so two light damages = 1 heavy damage and two heavy damages = destroyed, no matter when in the game this damage was received (and following this logic one light and one heavy damage = ¾ damaged - only requiring one furthe light damage to destroy it). A heavily damaged vehicle is neutralised and may not move or fire during its following turn.

8.50 If an AFV carrying external infantry passengers is destroyed by anti-tank fire a D6 is rolled to give the number of casualties.

8.51 For soft-skin vehicles (F class) hit at any range:

> 1, 2 or 3 = no effect.
>
> 4, 5 or 6 = the vehicle is destroyed.

8.52 When a soft-skin vehicle (F class) is destroyed by anti-tank fire the driver and any vehicle-mounted or towed weapons are lost.

Above: this element may suffer under rule 8.50.
Right: use of white smoke to show light damage to this Panzer IV.

High Explosive (HE) Firing

8.53 Weapons capable of firing high explosive rounds are guns (including many, but not all anti-tank guns), howitzers, mortars, multiple rocket launchers and infantry anti-tank launchers.

8.54 Mortars may fire twice per turn. Multiple rocket launchers fire twice but only on alternate turns. All other weapons fire once per turn.

8.55 Before determining any casualties a weapon firing HE must achieve a hit. The method for doing this depends on whether or not the weapon's fire is direct or indirect.

8.56 For each type of gun, the maximum range for direct fire and the table to use on the casualty chart to determine the effect of a hit is shown on the relevant VG chart.

Direct HE Firing

8.57 This is line-of-sight shooting of HE rounds when the firing element can see the target. **Direct Fire** is available to tank and self-propelled guns, deployed guns and tanks with HE capability.

> A 48" long length of thin wooden dowel marked with 8" bands is useful for measuring the majority of direct fire ranges.
>
> Playing Tip

8.58 To determine whether a hit has been achieved, divide the maximum range of the firing weapon into six equal bands. For example, a gun with a maximum range of 48" will have six bands each of 8". Band one = 0 - 8", band two = 8 - 16" and so on.

8.59 Measure the range from the firing weapon to the target (see **rule 8.5**) and calculate which band this falls within. The band number gives the basic D6 score required for a hit. For example: for a 25pdr field gun a target 22" away falls within band three (16 - 24").

8.60 Subtract 1 from this score if either the firer moved, or intends to move this turn, or if the target moved in its previous turn.

8.61 Once a hit is achieved from direct fire select the appropriate table on the casualty chart and throw a D6 to determine the effect on infantry and soft-skin vehicle targets.

8.62 If a direct fire target is in hard cover subtract 1 from the D6 score before determining any casualties.

Indirect HE Firing

IDF is carried out by weapons whose crews cannot observe the target because of intervening obstacles such as hills, or because the target is beyond the range of direct fire. The maximum range for IDF is 120" (72" for infantry guns and multiple rocket launchers). Deduct 24" from ranges if guns are off-table.

8.63 IDF is available to artillery pieces (including dual field and AT guns), multiple rocket launchers and 3" (or larger) mortars.

8.64 Guns normally fire IDF in batteries / battalions with no more than three guns. Some may have only one. For example, all five model guns in a 1944 Russian Light Field Artillery Regiment fire separately each turn.

8.65 Guns and mortars cannot fire IDF from inside a wood (i.e. more than 3" from the edge), or from jungle, unless deployed in a clearing at least 4" in diameter.

Artillery Observers (OPs)

8.66 IDF must be controlled by an observation post (OP). This is normally a radio-equipped two man team that may be on foot, on horseback, or in a vehicle. Other OPs include static posts equipped with a telephone, or artillery observation aircraft (AOPs). Each OP is normally attached to a specific artillery unit. AOPs are not attached and may control the fire of any artillery unit via a radio exchange vehicle (see **rule 8.70**).

8.67 OPs in vehicles or on foot may not move and control an artillery battery in the same turn. Telephone-equipped OPs may not move at all.

8.68 An OP may only select one target for IDF each turn. To do so, they must be able to observe the target and communicate with their artillery unit. If using radio, a D6 is thrown to test for contact each turn firing is attempted. A throw of 1 or 2 (only 1 for British, Commonwealth, American, late war Allied and German because of their greater number of radios) means no contact and, therefore, no firing that turn.

8.69 Due to their superior fire control methods British / Commonwealth and American OPs may direct the fire of one additional artillery unit, increasing, from 1943, to a maximum of six batteries (twelve guns) for a British and four battalions (twelve guns) for an American OP (this includes late war Allied units such as the Canadians, Poles and French). An OP can only control more than one artillery unit by radio if a radio exchange vehicle is deployed.

8.70 An OP attempting to control more than one artillery unit through a radio exchange vehicle must select a single target (which may not be changed) and nominate all the artillery units attempting to fire at it. Each of these artillery units must then take the radio test (see **rule 8.68**) before they can begin the IDF procedure. If an artillery unit fails the test, it may not use IDF during the turn and it cannot be directed to fire on another target by any other OP.

8.71 If both figures in an OP team become casualties the artillery unit they are observing for can no longer fire IDF. However, OPs allowed to control more than one artillery unit and AOPs may still attempt to direct their fire (see **rules 8.69 and 8.70**).

IDF Procedure

8.72 The diagram opposite page sets out the IDF procedure.

8.73 If a hit is achieved, select the appropriate table on the casualty chart (see **rule 8.7**) and throw a D6 to determine the effect.

8.74 If an IDF target is in hard cover subtract 1 from the dice score before determining the effect of the hit.

8.75 If the target does not move and it's OP passes a radio test, each gun of a battery that achieved an IDF hit on a stationary target in its side's previous turn can hit it again for a D6 throw of 4, 5 or 6.

ARTILLERY BATTERY FIRING PROCEDURE

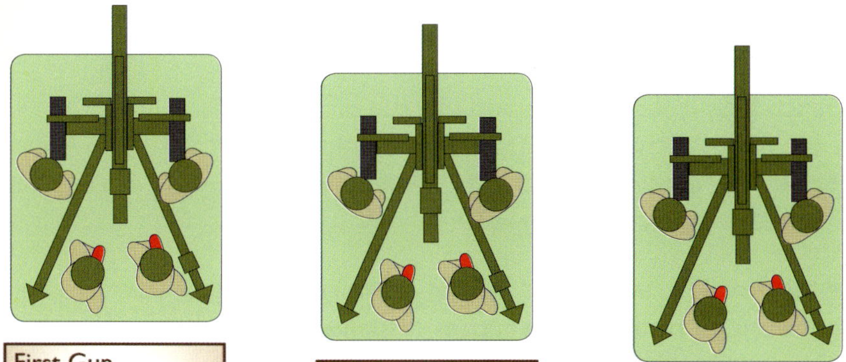

First Gun

Throw a D6. 5 or 6 is a hit (6 if the target moved in its previous turn)

If a hit, all guns in the battery are on target and throw for the effect of each hit

If a miss, go to the Second Gun

Second Gun

Throw a D6. 4, 5 or 6 is a hit (5 or 6 if the target moved in its previous turn)

If a hit, this and the third gun (if there is one) are on target

If a miss, go to the Third Gun

Third Gun

Throw a D6. 4, 5 or 6 is a hit (5 or 6 if the target moved in its previous turn)

If a hit, throw for the effect of one gun only

Note: many artillery batteries / battalions are two gun units

An infantry battalion OP.

OPs in Infantry Battalions

8.76 Mortars and guns that are part of an infantry battalion or regiment do not require radio tests for IDF. Any figure in the HQ group of the battalion, regiment or brigade may act as an OP for these weapons and can direct their fire if they can see the target. Battalion weapons can only be controlled by their own HQ group and not by HQ figures from other battalions. These OPs <u>do not</u> add 12" to their observation ranges.

8.77 If one of the battalion's companies is defending an isolated position, any HQ figure may be attached to act as an OP for the battalion's support weapons.

HE Effect

Hits on Infantry, Cavalry, Support Weapons and Deployed Guns

8.78 Each HE hit creates a 4" diameter risk zone which is centred on the aiming point on the target (see **rule 8. 3**). Any figure or weapon in the risk zone is a potential casualty.

8.79 A hit on a vehicle or building / structure will only cause casualties or damage to that vehicle or structure or elements in or on it. There is no effect on adjacent figures and weapons (see **rules 8.83 to 8.87**).

8.80 For each HE hit throw a D6 and use the HE value (detailed in the VG chart) and the casualty chart to determine the number of casualties. A further D6 is thrown for each gun or infantry support weapon in the 4" risk zone. A 6 means the weapon is destroyed.

8.81 The test for destruction of a gun or support weapon should also be taken if an autocannon or a grenade causes casualties amongst the crew. If no casualties are lost then no test is necessary.

8.82 Small arms fire cannot damage any gun or support weapon.

HE Hits on Soft-Skin Vehicles

8.83 If a soft-skin vehicle receives two casualties 'worth' of damage in a turn it is destroyed and the driver is killed. Any additional casualties inflicted during the same turn are removed from the vehicle's passengers. If one or more of these casualties are caused by HE fire any mounted or towed weapon is lost and a further D6 is thrown for each support weapon (such as an MMG, mortar or AT rifle) carried in the vehicle. 6 means the weapon is destroyed.

8.84 If a soft-skin vehicle only sustains one casualty 'worth' of damage from HE or small arms fire during a turn there is no effect on the vehicle. This casualty is not carried forward to the next turn.

HE Hits on Armoured Vehicles (AFVs)

8.85 When an open-topped AFV receives a hit from IDF (including any mortar fire) throw a D6. 6 means the vehicle is destroyed and any towed weapon is lost. Direct HE fire is not permitted against AFVs. AT values (as detailed on a VG chart) and the anti-tank firing procedure set out in **rules 8.39 to 8.42** and **Table 10** should be used.

8.86 A hit from IDF may destroy a closed AFV (such as a tank or fully-enclosed assault gun) if the calibre of the gun firing is 80mm or over. The target vehicle can only be destroyed by a D6 throw of 6.

8.87 If an AFV carrying external infantry passengers (such as tank riders) is destroyed by HE a further D6 is rolled to give the number of casualties. Any result under 6 may cause casualties - consult the relevant table on the casualty chart and treat the passengers as being in the open.

HE Hits on Buildings and Bridges

8.88 Before the game begins, decide which buildings are intact and which are in ruins. When calculating HE fire effect against elements in intact buildings, subtract 1 for hard cover as normal, but don't remove any figures for the first three casualties. After three casualties 'worth' of damage the building is reduced to a ruin and any surviving occupying figures are then considered to be in hard cover for further HE hits. Casualties are then removed as normal

8.89 Wooden houses only offer soft cover (unless agreed by all players to be fortified before the game with sandbags, concrete or internal trenches). They are completely destroyed after two casualties 'worth' of damage. Fortified wooden buildings are treated as intact (hard cover) buildings.

8.90 Bridges are weakened after ten casualties 'worth' of damage and destroyed after 15. Weakened bridges may only be crossed by infantry. Other structure values must be agreed before the game.

Multiple Rocket Launchers

8.91 These include weapons with multiple barrels (such as Nebelwerfers) or multiple launching rails (such as Katyushas). Two shots 'worth' of fire each turn are allowed to simulate the heavy fire power of these weapons. Because multiple rocket launchers were less accurate than guns the fire grid is used to plot hits.

- The centre square (5) of the 12" clear plastic fire grid (divided into nine numbered 4" x 4" squares) is placed over the target, pointing in the direction of fire.
- Throw a D10 to give the number of the square where the shots lands. Anything within this square is a hit. 0 is a miss.

8.92 Throw twice for effect (D6) against targets in the square. Targets must be wholly within the square and players may position the grid so more than one square straddles a target.

1	2	3
4	Target 5	6
7	8	9

Support Weapons

Light Mortars

8.93 Light mortars are those with a calibre of less than 3" or 81mm.

8.94 Light mortars can fire HE and smoke and have a minimum range of 3" and maximum range of 18". They may fire twice (at the same target only) if stationary and once if moved. They hit their selected target for a D6 throw of 4, 5 or 6.

Below: a British 2" light mortar.

Above: a German 81mm medium mortar.

Below: a British 4.2" heavy mortar.

Medium and Heavy Mortars

8.95 Medium and heavy mortars can fire two rounds of HE or smoke per move at the same target. They may not move and fire in the same turn.

8.96 Even if organised in batteries, medium and heavy mortars always test individually for hits. They are always treated as firing IDF, even if the mortar crew is able to observe the target. For example: the crew of a 4.2" mortar can observe their target and don't need an OP team. However, they still have to test to hit as for IDF.

8.97 Medium and heavy mortars need a 5 or 6 on a D6 throw to hit with their first round (6 if the target has moved in its previous turn) and this gives an automatic second round hit (on the same target). If the first round is a miss the second round needs a 4, 5 or 6 (or a 5 or 6 if the target is moving) to hit.

Table 11: Medium and Heavy Mortars

Mortar	Points Table for HE	Ranges
Medium (3" / 81mm)	8 points on table 4	Minimum 6" Maximum 48"
Heavy (4.2" / 120mm)	12 points on table 4	Minimum 6" Maximum 60"

Hand-Held Anti-Tank Weapons

8.98 Hand-held infantry anti-tank weapons fall into two categories: anti-tank launchers and anti-tank rifles. Anti-tank rifles fire at three ranges:

Short 0 - 6", medium 6 - 12" and long 12 – 18".

Anti-tank launchers (such as the PIAT, Bazooka, Panzerfaust and Panzerschrecke) have a single range of 6".

8.99 Firing procedure is as for AT guns (see **rule 8.42** and **Table 10**). Anti-tank rifles have a gun classification of 6. Anti-tank launchers have a gun classification of 2.

8.100 Only anti-tank rifles, PIATs and Panzerfausts may be fired from intact buildings.

8.101 None of these weapons may be fired at infantry in the open or soft cover. Anti-tank launchers may be fired at any hard cover target with an effect of 8 points on table 4.

8.102 Panzerfausts may only fire once in a game. Other anti-tank launchers may fire three rounds per game. Anti-tank rifles have unlimited ammunition.

Flamethrowers

8.103 Maximum range is 6" for man-pack and 9" for vehicle flamethrowers.

8.104 Man-pack flamethrowers are carried by a single figure and have a firing value of 15 points on the casualty chart.

8.105 All flamethrowers have an anti-tank capability. Fully-enclosed armoured vehicles are destroyed for a D6 throw of 3, 4, 5 or 6. Open-topped AFVs or soft-skin vehicles are destroyed for any throw but 1.

8.106 If any deployed gun or support weapon is fired at by a flamethrower throw another D6. The target is destroyed for any throw other than 6.

8.107 Buildings (but not bunkers or ruins) hit by flamethrowers will burn for three moves. Figures inside suffer no casualties, but have two full moves to evacuate the building or will automatically be lost. Very large buildings (such as factories) may be treated as a number of buildings depending on their size. Throw a D6 to see if a 'section' next to the burning part of the factory catches fire. 5 or 6 means it also burns.

8.108 Troops inside a burning building may not fire.

8.109 Flamethrowers can set fire to wooden bridges for a D6 throw of more than 1. Bridges are impassable when burning (for three moves) and are then destroyed.

Final Movement

9.1 Any element that has not already moved during this turn can now be moved.

9.2 Some weapons cannot use the **Final Movement** option if they have fired during this turn. These are:

- MMGs and HMGs.
- Medium and heavy mortars and light mortars that have fired twice.
- Guns that have fired IDF.
- AT guns on cruciform mounts (such as the British 2pdr and German Flak 88mm).

9.3 No final movement can be used to bring a unit into base-to-base contact with an enemy unit and / or a close assault.

88mm guns, HMGs and medium mortars may not move and fire in the same turn.

Action at Lingevres

Top: A Company in the cornfield (move 2).
Centre: 1st Company dug-in in the wood (move 1).
Bottom left: Sherman attacked by Panzerfaust (move 5).
Bottom right: The British defend the village (move 9).

Lingevres - A Battalion-Level Scenario Played with the Basic Rules

The British attack a German-held village in Normandy, 14th June 1944

Background

It's eight days after D-Day. The 50th (Tyne Tees) Division is part of the British 21st Army Group's attempt to drive deeper inland, but blocking their path is Panzer Lehr, one of the German Army's strongest armoured divisions. The British division has been given the task of breaking through the German defences and, as part of its 151st Brigade, the 9th Durham Light Infantry have been ordered to capture the small village of Lingevres. Panzer Lehr's Panzer Grenadier Regiment 902 is responsible for defending the village and elements of its 2nd Battalion are dug-in on the outskirts and amongst the buildings, with tanks of Panzer Regiment 130, in support.

Objectives

The British win if they capture and hold the village within twelve moves. The German player must prevent this to win. The game is a draw if both sides end the game with their forces still active in the village, or if both sides are forced to withdraw. The game is played with the **Basic Rules**.

Forces

All units British units are regular. All German units are elite.

Panzer Grenadier Regt 902
(Oberstleutnant Welcsh)

2nd Battalion HQ
CO + 4 figures, Panzerschrecke
Sdkfz 251/1 half-track

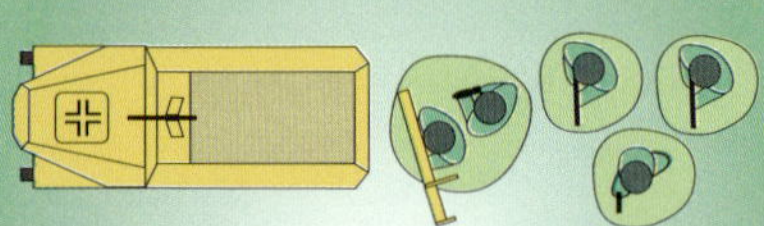

Company
8 figures, Panzerfaust

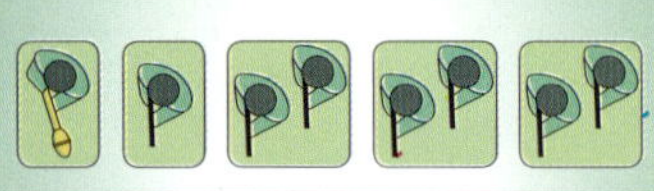

Company
8 figures, Panzerfaust

Heavy Company
3 figures, MMG
3 figures, 120mm mortar, Sdkfz 251/1 half-track

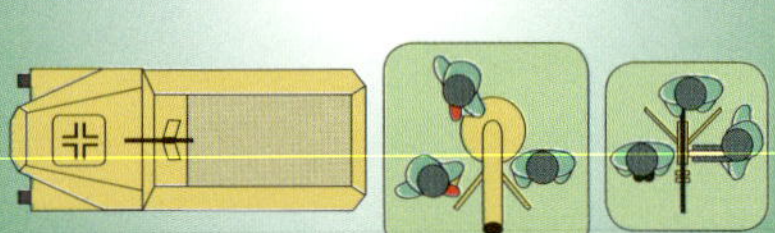

2nd Company, Panzer Regiment 130
Panther (75mm)

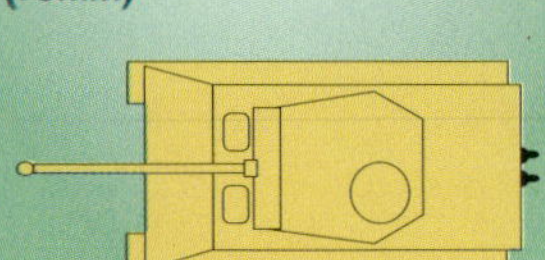

German Reinforcements (From Move 8)

Company from 1st Battalion
8 figures, Panzerfaust
Sdkfz 251/1 half-track

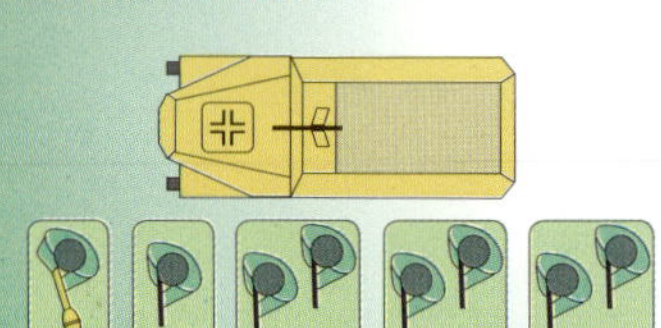

2nd Company, Panzer Regiment 130
Panther (75mm) (Lt Steindamm)

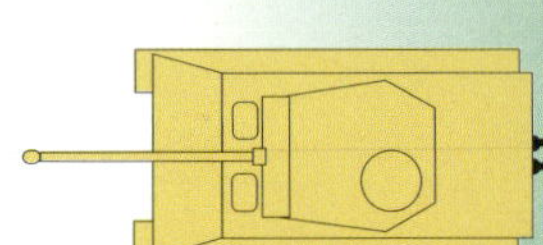

Vehicle and Gun Chart - Lingevres

Vehicle (Main Weapon)	Speed	Armour	Hull Weapons	Turret Weapons
Panther tank (75mm)	30" / 15"	B	VMG	AT class 1 / HE 8 points table 4 (or VMG)
Sherman tank (75mm)	24" / 12"	C	VMG	AT class 3 / HE 10 points table 4 (or VMG)
Sherman Firefly tank (17pdr)	24" / 12"	C	-	AT class 1 (or VMG)
Achilles tank destroyer (17pdr)	24" / 12"	D		AT class 1 (or VMG)
Sherman OP tank	24" / 12"	C	VMG	VMG
Sdkfz 251/1 half-track	30" / 15"	E	LMG	
Bren Carrier	30" / 15"	E	LMG	
Bren Carrier (Vickers MMG)	30" / 15"	E	MMG	
Loyd Carrier	24" / 12"	F	-	
Gun				
6pdr anti-tank gun	AT class 3			

The opposing HQs.
Left: Oberstleutnant Welcsh in Berolles.
Right: Lt Colonel Wood in front of the OP tank.

151st Infantry Brigade

9th Durham Light Infantry
HQ (Lt Colonel Wood)
CO + 6 figures, PIAT, 2" mortar

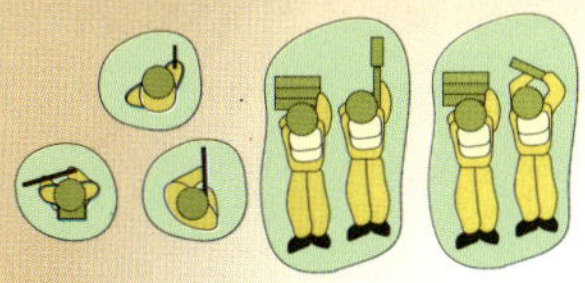

Carrier Platoon
3 figures, PIAT, Bren Carrier

A Company
8 figures

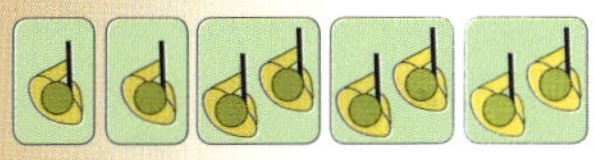

B Company
8 figures

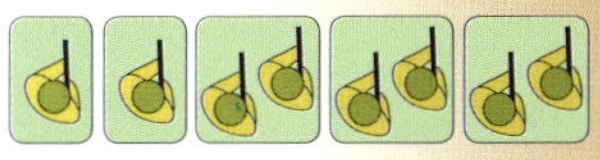

C Company
8 figures

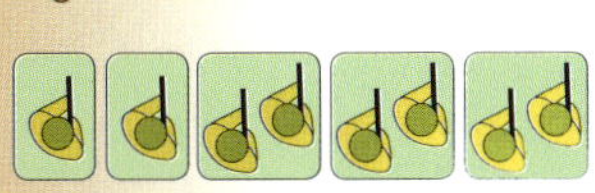

D Company
8 figures

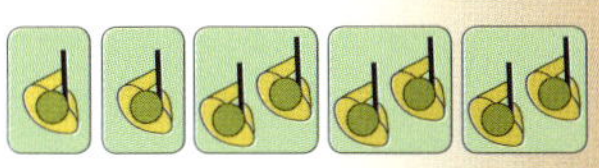

Support Company
3 figures, 6pdr anti-tank gun, Bren Carrier
3 figures, 3" mortar, Bren Carrier

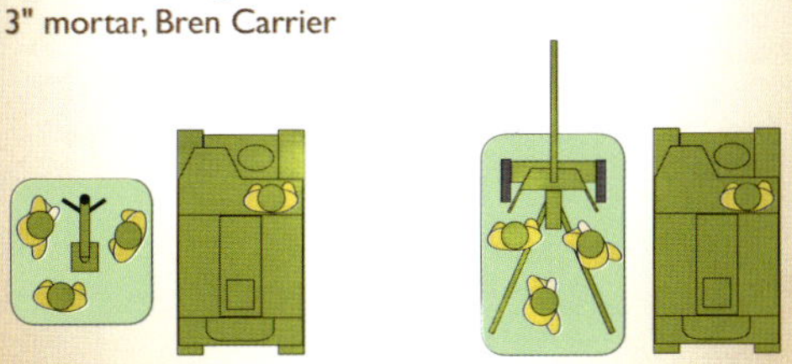

MG Company, 2nd Cheshire Regiment
CO + 3 figures, PIAT, Bren Carrier
3 figures, Vickers MMG, Bren Carrier
3 figures, 4.2" mortar, Loyd Carrier

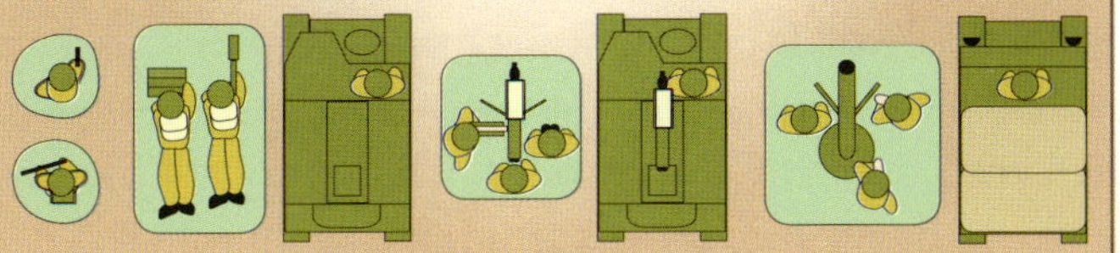

Supporting Units

A Squadron,
4th/7th Dragoon Guards_
(Major d'Avigor-Goldsmid)
2 x Sherman (75mm)
Sherman Firefly (17pdr)

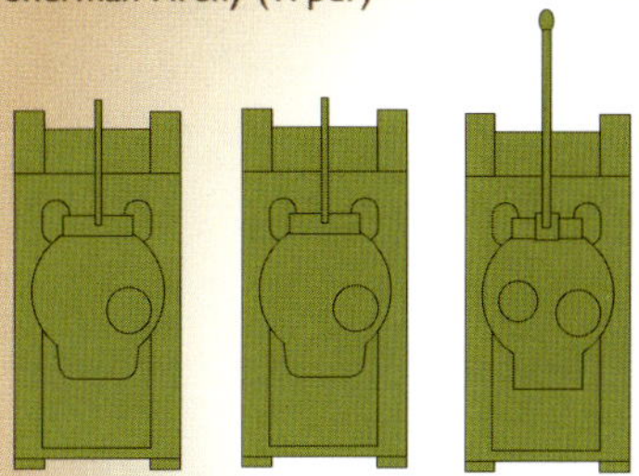

Troop,
102nd AT Regiment
Achilles (17pdr)

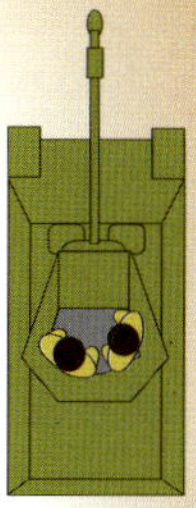

341st Battery
86th Field Regiment Royal Artillery
2 x 25pdr field gun
(Off table)

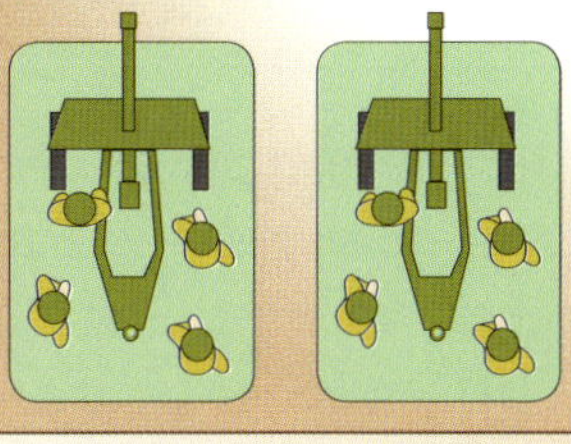

Battery OP
Sherman OP tank
(On table)

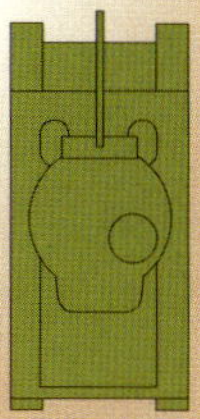

In our play test the British player decides to attack on a broad front. The infantry are deployed with two companies (A and B) forward and two behind (C and D) in extended lines; a typical British formation. The tanks, OP and other supporting units are in the centre and on the flanks. The mortars are deployed.

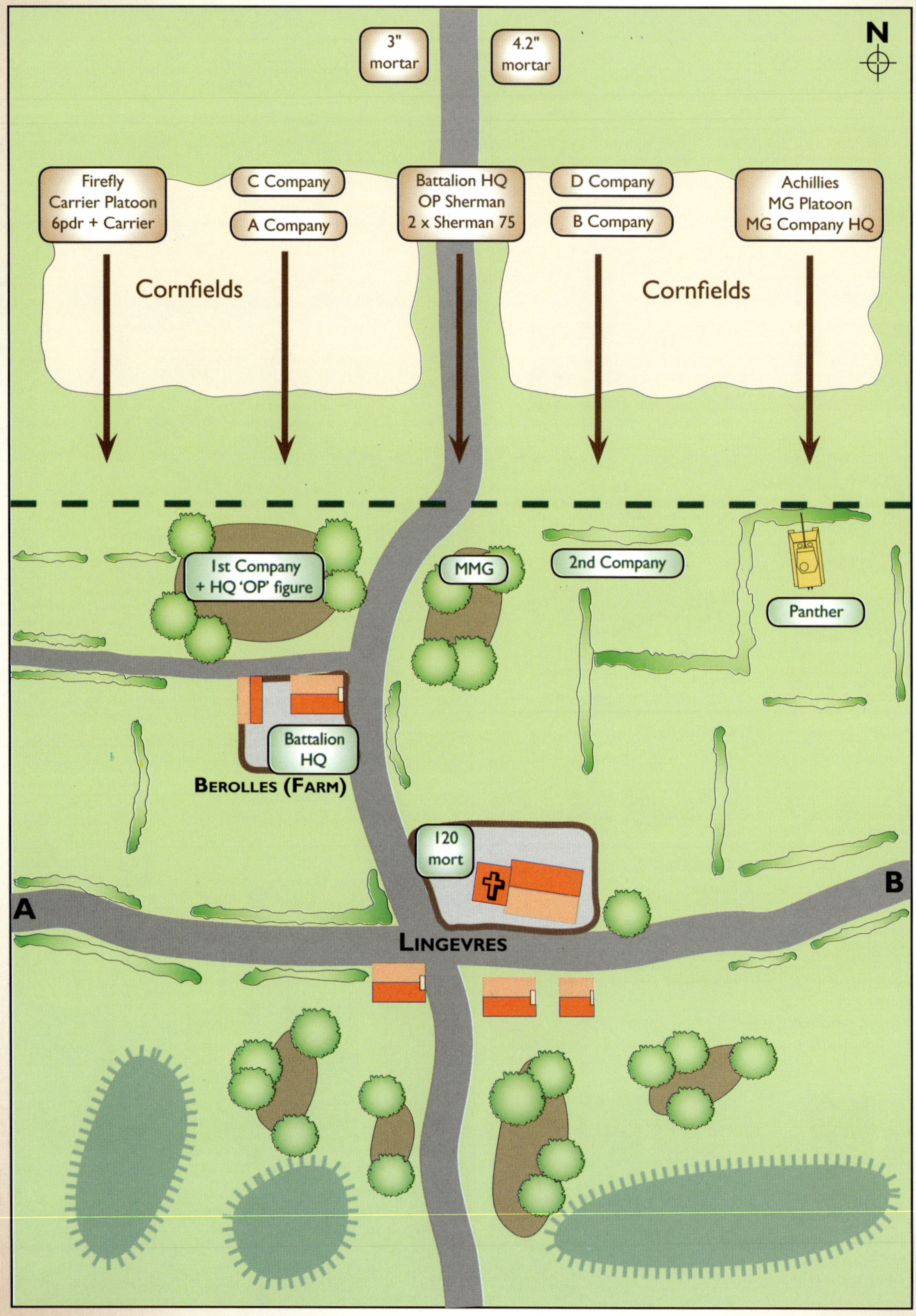

For this game the German figures and models are placed on the table, even though they all start the game in cover. Alternatively, they could have been replaced with markers, placing the elements on table when observed.

Although a command map (**advanced rule 10.1**) is not essential when playing with the basic rules, a quick sketch plan and / or a few notes will help your memory during the game.......

The German player decides to deploy well forward, hoping to hit the British as they cross open ground. Panzer grenadiers are dug-in in the woods (1st Company) and behind a hedge (2nd Company), with their HQ in Berolles. The lone Panther tank is behind cover on the right and the heavy mortar is set up in the churchyard.

The Game Diary

Who moves first? A D6 throw of 5 (as opposed to the German throw of 2) means the British take the first turn in move 1. Let battle commence!

Move 1

British Turn	
Morale	X
Observation	The OP tests to observe the Panther concealed behind the hedgerow. The normal range for a chance test is 24" which puts the tank out of range, but an OP can add another 12". The D6 throw is 2, not the 5 or 6 required, so the Panther stays hidden!
Smoke	X
Heroic Actions	X
Movement	The infantry advance through the cornfields at their maximum 6" cross country move with no deductions. The tanks keep pace.
Reserved Fire	Confident of a kill at normal range, the Panther opens fire at the Achilles. Its class 1 gun is pitted against D armour, needing just a 2 to hit, but 1 is deducted from the D6 throw because the Achilles has moved. The German player throws a 1 and misses!
Close Assault	X
Firing	Having fired, the Panther can now be observed. The Achilles fires its class 1 gun against the Panther's B armour, needing 3. The D6 throw is 5, -1 because the Achilles moved and -1 because the target is partially concealed. It's a hit! Another throw and a 4 means the Panther is heavily damaged and can't fire or move in its next turn.
Movement	X

German Turn	
Morale	The heavy damage means the Panther's crew have to test for morale. Their elite status (+1) cancels out the damage (-1) and with a throw of 6 they can stay with their immobile tank.
Observation	The OP in the wood can automatically observe the British HQ
Smoke	X
Heroic Actions	X
Movement	X
Reserved Fire	All eligible weapons have moved or fired, so there's no reserved fire.
Close Assault	X
Firing	With no need for a radio test, the battalion's 120mm mortar opens fire on the British HQ. As the target moved in its last turn the mortar needs a 6 to hit with the first round, but throws a 3. With the second round a hit can be achieved for 5 or 6, but a throw of 2 means another miss.
Movement	X

Move 2

British Turn	
Observation	The OP Sherman advanced during the previous turn's Movement stage, so can't contact the artillery battery this turn to direct fire.
Movement	The infantry and tanks (except the OP) advance 6". The Achilles moves to within short (AT) range of the damaged Panther. The Carrier Platoon scouts ahead.
Reserved Fire	The German MMG opens fire, its 8 points directed at A Company in the open at medium range. A throw of 3 means the Durhams lose their first casualty: one figure.
Firing	Now within short range (+1), the static Achilles opens up on the disabled Panther. It is still partially concealed and the Achilles only manages a throw of 2: a miss, leaving the British SP gun very exposed. Having fired, the MMG is targeted by both 75mm Shermans. It is within range band 2 (8 -16"), and the tanks deduct 1 from D6 throws for moving. One tank hits. A 75mm gun has an effect of 10 points on table 4, but even with -1 deducted for hard cover, a 6 inflicts three casualties, wiping out the MMG crew. The gun survives its test for an HE hit.

German Turn	
Observation	The OP in the wood automatically observes B Company advancing to its right.
Movement	As it is a battalion support weapon the German player can re-crew the MMG with infantry from 2nd Company. Three figures are within 4", but the wood is a continuous obstacle. A throw of 2 (-2" from their 6" move) allows the move.
Firing	1st Company opens fire. Seven figures, + 2 for their elite status, firing at medium range at A Company in the open totals nine points. 3 on a D6 means the British lose one figure. The MMG again targets A company, as does 2nd Company behind the hedge. The result: four British casualties. The Panther, now restored (but with HD noted), fires at the Achilles and hits, causing heavy damage. Finally, the 120mm mortar targets B Company. The first throw misses, but the second (a 5) is a hit. At 12 points on table 4, a score of 3 inflicts three British casualties.

Move 3

British Turn	
Morale	Total losses are nine figures; nowhere near 20 or half battalion strength. The IDF fire was from mortars, not guns over 80mm, so no tests are required. The heavily damaged Achilles does need to test and throws a 4: no effect.
Observation	Having fired, both enemy companies and the MMG can be observed.
Smoke	The OP tank contacts and directs the artillery against the Panther. Both rounds hit, giving a 6" wide smoke screen in front of the tank to mask the helpless Achilles.
Movement	The Carrier Platoon dash to the edge of the wood and debus. The tanks and survivors of A and B Companies advance to within 6" of the defenders. The MG Company HQ's PIAT team debus to stalk the Panther.
Firing	A and B Companies fire at short range and tank HE and VMG fire cuts into the defenders, causing a shattering twelve casualties in one turn.

German Turn	
Morale	The German battalion has now lost 15 figures: half its strength. But as the majority are occupying defences (+1) and are elite (+1) a throw of 3 means there is no effect.
Movement	The few survivors are now pulled back, 1st Company in the wood slowed by D6 deductions. The Panther shifts position to avoid the smoke but is tempted to target one of the Shermans. At 19" range it hits, but only causes light damage.

Above: the carrier platoon in move 3. Below: The Panther masked by the smokescreen in move 3.

Move 4

British Turn	
Movement / Firing	The restored Achilles (HD noted) and MG Company flank the Panther, as A and B companies reach the edge of the wood and hedge with tanks in support. On the right, the Firefly and remounted Carrier Platoon drive forward. The Achilles and nearest Sherman miss the Panther, but B Company and a Sherman VMG wipe out the 2nd Company and MMG crew pulling back in the open.

German Turn	
Morale	With four more casualties in the last turn, the German battalion must test again. +1 for elite is cancelled out by -1 for a consecutive test. With a paltry throw of 2 the battalion is pinned down. 'Pinned' units in the open must withdraw a full move towards friendly cover, but 1st Company is half out of the wood and half in the open, already heading towards Berolles. It is logical to continue.
Movement / Firing	The Panther misses the Achilles at point blank range (a throw of 1...again!), then turns and heads back to the village.

Move 5

British Turn	
Movement / Firing	The Achilles / MG Company on the left and the Firefly / carriers on the right are aiming to seal off the two German entry points. The infantry make slower progress moving through the wood and hedgerows, but the Shermans push on along the road. The leading tank fires at the rear of the retreating Panther, causing light damage. The Firefly also spots the Panther and the shot causes heavy damage. Added to the heavy damage suffered on move 1 this destroys the German tank.

German Turn	
Morale / Movement / Firing	With no further losses, the infantry survivors in Berolles don't need a test. Wasting no time, an element with a Panzerfaust moves to cover the road. It targets the lead 75mm Sherman, which is hit and destroyed.

Move 6

British Turn	
Morale	A total modifier of -2 (tank destroyed / enemy in hard cover within 12") is countered by a throw of 5. The tank squadron is ok
Movement	The Sherman on the road pulls back as the Achilles and MG carriers reach the eastern edge of Lingevres. On the right, A and C Companies move swiftly through the wood (a 1" deduction) while the Firefly and carriers near the road into the village. The mortars pack up and move forward.
Firing	The Sherman OP successfully calls in IDF on Berolles. Two hits cause four casualties amongst the enemy survivors and the German CO becomes a casualty.

German Turn

Morale	With casualties lost in the last turn the throw is 4. Modifiers cancel each other out: -1 (no friends within 12"), -1 (incoming IDF over 80mm), +1 (elite), +1 (hard cover). The battalion morale is ok.

Move 7

British Turn

Firing	Small arms fire into Berolles is ineffective, but the PIAT causes two casualties.

German Turn

Morale	With only four figures (+ the three mortar crew) surely the battalion is finished. The throw is 5, the modifiers: -1 (no CO), -1 (second consecutive test), -1 (no support within 12"), -1 (under artillery fire over 80mm), +1 (hard cover), +1 (elite). The result: 5 – 2 = 3. No offensive action. Although on its last legs, the battalion is still buying valuable time for the reinforcements by tying down half of the British infantry.

Move 8

British Turn

Close Assault	A and C Companies close assault Berolles. Twelve attackers + a D6 throw of 5 = 17. Four defenders + a D10 throw of 7 = 11. With a clear 2 points difference between the scores the assault succeeds, the Germans losing three figures and the British one.
Firing	The Sherman hits the 120mm mortar and wipes out the crew.

The initial German force has been destroyed, for a cost to the British of nine casualties and one tank destroyed.

German Turn

With the game now hinging on the German reinforcements, they throw for their arrival point and enter at A. Reserved fire from the Firefly and the 6pdr miss the Panther and it returns fire, inflicting light damage on the Firefly, which is partially concealed behind a house.

Move 9

British Turn

The church is finally occupied by B and D Companies and the Sherman covers the crossroads. The Firefly fires at the Panther and misses (again), but the 6pdr inflicts light damage.

German Turn

The Panther swings south through a gap in the hedge and out flanks the British positions, with the 3rd Company half-track following. The Panther can't claim a rear attack, but still hits and destroys the Firefly!

Move 10

British Turn

The tank squadron tests for morale, resulting in 'no offensive action'. The Achilles is now the only mobile defence against the Panther and advances to a firing position. The debussed MG Company PIAT also moves towards the tank. A and C Companies and the OP reach the hedge along the western road having cleared Berolles. The Achilles fires at and hits the Panther, causing heavy damage.

German Turn

If they debuss, 3rd Company in the half-tack will be vulnerable to fire from a nearby house, but they are the Panther's only protection against the MG Company's PIAT (which needs a Heroic Action test to attack). They decide to stay put and use reserved fire if necessary.

Move 11

British Turn

The PIAT team's Heroic Action test fails and the Achilles misses the Panther! However, the OP Sherman contacts its guns and they score two hits and destroy the half-track.

German Turn

With the half-track destroyed and the Panther carrying a heavy and light damage penalty, the German commander realises the game is up. The Panther reverses and 3rd Company's survivors pull back. For them the game is over...

Debrief

The British won quite clearly, although the second Panther caused a lot of damage and might have tipped the balance if the initial defenders could have held on for a little longer. You may think the German player made a bad mistake defending so far forward. Could a more 'hit and run' approach have been more effective and was it wise for the Panther to be in the front line? Nonetheless, the British tactics were rather predictable and, although typical of the period, are extended lines really the best way to attack a dug-in defence? Asking these and many other questions will hone your skills as a table-top commander. Remember, the best generals learn more from defeats than victories.

Advanced Rules

The real battle for Lingevres was simplified for our re-fight. Aspects of the battle, such as a 'pre-game' artillery bombardment and air attacks were omitted, but can be added using the Advanced rules. If players decide to follow this suggestion, add a 75mm Pak 40 anti-tank gun and a 'Drilling' triple 15mm anti-aircraft half-track to the German force to balance the scenario.

Advanced Rules

After playing a few games with the **Basic Rules**, most players will want to expand their games to include other important elements of WWII warfare, such as engineering and the use of aircraft. The **Advanced Rules** cover these and much more, as well as adding detail and options to the basic principles of play.

The Advanced Rules sections are listed below alongside the **Order of Play** stages most commonly linked to their rules. For example, in the **Fighting in Built-up Areas** section the majority of the rules cover **Movement** and **Firing**.

Advanced Rule Section	Order of Play Stage(s)
10 **Command and Control**	Pre-Game and Morale
11 **Supply, Repair and Medical Support**	Morale and Final Movement
12 **Engineering and River Crossings**	Movement and Firing
13 **Specialist Small Arms Fire**	Reserved Fire and Firing
14 **Specialist Artillery Fire**	Smoke and Firing
15 **Aircraft**	Movement and Firing
16 **Fighting in Built-Up Areas**	Movement and Firing
17 **Paratroop and Glider Landings**	Movement
18 **Amphibious Landings and Naval Forces**	Morale, Movement and Firing
19 **Naval Combat**	Movement and Firing
20 **Terrain, Climate and Darkness**	Observation and Movement

An American 'higher command' unit.

Maps

10.1 Before the game, each player must draw a simple command map of the battlefield as set up on the playing table. Orientate the map by choosing and marking on a direction for north, then draw:

For attackers	Arrowed lines for each major unit's planned route of advance and rings around units' objective(s).
For defenders	The positions of any concealed units (if not using markers), unmarked minefields and prepared defensive positions not obvious on the table.

Orders

10.2 Players must give each of their major units simple, one-sentence long written orders before the game begins. For example:

Infantry Battalion -	*Advance to capture wood A and, if successful, advance to capture hill B.*
Cavalry Regiment -	*Move mounted on the northern slopes of ridge C, then dismount to dig in and defend the ridge.*
Tank Battalion -	*Advance on the far left flank in support of battalion D to capture village E.*
Artillery Battery -	*Carry out observed defensive fire against any attacking enemy units.*
Recce Battalion -	*Advance to cross-roads F. Retreat if fired upon.*

10.3 The orders should also include the move on which any pre-ordered aircraft strikes are due to take place (see **rule 15.3**).

10.4 Units must follow their orders as closely as possible, allowing for the effects of enemy actions and morale tests.

10.5 Major units are normally battalion-size units, but may include tank, reconnaissance and engineer companies and artillery, anti-tank or AA batteries / battalions. However, in very large games major units may be upgraded to the next level of organisation (i.e. from infantry battalion to brigade, or artillery battery to regiment).

Change of Order Test

10.6 Players may test to significantly change any of their major units' orders as long as the highest level HQ is still functioning (see **rule 10. 5**) and any of the following circumstances apply:

- A unit has achieved its initial objective.
- An attacking unit has been under fire and unable to advance for two full moves.
- A unit is in danger of being over run, outflanked or surrounded.
- A unit is suppressed, pinned down, or has been forced to take no offensive action or retire by a morale test.
- The player wishes to redeploy the unit to reinforce a successful attack or a threatened defence.

10.7 Regular units (depending on nationality) need the D6 throws shown in **Table 12** if they wish to adopt new orders. Carry this out after any compulsory morale tests.

TABLE 12; CHANGING ORDERS

German, British, Commonwealth, US and other late war Allies.	3, 4, 5 or 6
Japanese, 1940-41 Finnish, 1940 Polish.	4, 5 or 6
Soviet, Italian, Romanian, Hungarian, Slovak, 1942+ Finnish, Chinese and early-war French, Dutch and Belgian.	5 or 6
Elite units add 1 to their throw. Poor units and units reacting to a morale test subtract 1 from their throw.	

10.8 If a unit fails a test it must continue to follow its original orders. Units that fail may test again in the following moves.

Highest Level HQs

10.9 Before the game begins each player must identify their highest level HQ. In most games this will be a regimental / brigade headquarters, or a combined arms unit HQ, such as a US Combat Command or German kampfgruppe. Highest level HQs are represented by six figures and one or two command vehicles.

10.10 If the player's highest level HQ suffers 100% casualties, or is captured or dispersed by an enemy attack, any major unit under its command deducts 1 from morale throws and may not test for a change of orders.

10.11 In large games, a side using more than one regiment / brigade may field a divisional HQ. If one of the regimental / brigade level HQs is destroyed all major units under its command deduct 1 from any morale throws as usual. However, at least one full move after this HQ has been eliminated, its major units may test for a change of orders from the divisional HQ (i.e. if a brigade HQ is destroyed on move 9 its major units may not test for a change of orders from divisional HQ until move 11).

> Divisional HQs can be identical to other highest level HQs, but players may choose to add interesting extras such as a military police element or a despatch rider.
>
> **Playing Tip**

10.12 COs of highest level HQs may attempt to rally routers by moving to within 3" of the unit concerned. Throw a D6. A score of 6 (5 or 6 if elite) means the unit will stop routing, may move up to 6" into cover if desired and must spend two full moves recovering before further action. If the unit suffers any casualties whilst recovering, throw for morale as usual.

Supply

11.1 Players may agree to restrict tank ammunition to five rounds per vehicle to introduce the problems of re-supply.

11.2 To replenish a tank, a supply vehicle must be in contact for a full move, starting during final movement, with the following restrictions.

- Up to two tanks may be re-supplied at the same time by one vehicle.
- One supply vehicle (medium truck, half-track or a converted light tank) carries enough ammunition to supply a tank battalion once in a game.
- The supply vehicle itself may be replenished after one move in contact with a 'full' supply truck or an on-table ammunition dump.

Vehicle Repairs

11.3 A specialist repair vehicle is needed to repair damaged AFVs. It must be in contact with the damaged vehicle and takes one full move to remove one heavy damage or two light damages. Two AFVs with light damage can be repaired at the same time by one repair vehicle.

11.4 If an AFV with heavy damage cannot be repaired in position (e.g. due to enemy fire) it may be towed to a safer location by a recovery AFV or heavy tractor (which may, or may not be the repair vehicle) at no more than 6" per move. Deduct 3" for connecting the towing cable.

11.5 Repairs and vehicle recovery take place during final movement.

Medical Support

11.6 Each battalion-size infantry / cavalry unit adds a 'medic' figure that may be attached to any sub-unit. If the medic is present, the sub-unit may attempt to 'save' each casualty suffered that turn by throwing a D6. If the score is 5 or 6 they survive. All other casualties are then assumed to have been taken to the battalion HQ.The medic cannot become a casualty.

11.7 Each brigade / regimental HQ adds an advanced dressing station (ADS) and a light medical vehicle that, during final movement, may evacuate up to three casualties (still counted as morale losses) at a time, from a battalion HQ to the ADS, with a 6" embuss / debuss deduction. When there are ten casualties in the ADS throw a D6 to give the number of figures to be returned to the battalion(s) next turn. This may not exceed a battalion's total casualties and no CO figures or other specialist figures may be returned to the unit. The rest stay in the ADS until the total reaches ten and another test is taken.

11.8 An ADS takes one full move to dismantle / set up and moves with the regimental / brigade HQ. If the enemy overrun the ADS no more casualty evacuation may take place for the rest of the game.

Repair Vehicles, Supply Trucks and Medics

To help keep track of ammunition use identify individual tank models by painting a squadron or company letter and tank number under the vehicle or its base (e.g. the second Sherman in C Company will be 'C2').

Light medical vehicles can be modified Jeeps or light ambulances, including horse drawn vehicles. Represent the ADS with a tent, medical figures and a Red Cross flag.

To keep track of casualties in the ADS, move figures evacuated from a battalion into one corner of the off-table tray used to hold casualties. If returned to action, place them back with the unit.

Playing Tips

Top: an American ARV (armoured recovery vehicle) tows a Sherman.
Centre left: a German Maultier supply truck in action.
Centre right: a medic figure marker.
Bottom: a German Famo repair / recovery half-track at work.

12.1 All engineering tasks are carried out in the first movement stage of a turn, except for demolitions and petard and flame attacks which are carried out in the firing stage. Any move of engineering activity takes the entire first movement stage of a turn and no other movement or engineering work is permitted.

Anti-Tank and Infantry Obstacles

12.2 Obstacles include:

- Barbed wire and improvised obstacles to vehicle movement, such as vehicle road blocks, paving stone barriers, felled trees and road craters.
- Permanent obstacles to vehicles, such as concrete blocks, anti-tank walls, girder barriers, emplaced tree trunks and anti-tank ditches

12.3 Barbed wire and improvised obstacles may be constructed before or during a game. Permanent obstacles may only be constructed before a game.

12.4 A 4" length of barbed wire, or a 4" wide improvised obstacle to vehicles is prepared by a minimum of two engineers or four infantrymen in two moves. There must be material for the obstacles nearby. An engineer can blast a crater the width of a road after two moves of preparation.

12.5 Barbed wire delays infantry and wheeled vehicles for one move once they are in contact. Tracked AFVs test as for crossing linear obstacles (see **rule 5.4**) and make a 4" wide gap.

12.6 Improvised obstacles are impassable to wheeled vehicles and are linear obstacles to tracked vehicles and infantry. They are cleared by a minimum of two engineers or four infantry figures in two moves. A bulldozer or tank dozer can clear an improvised obstacle in one move.

12.7 Road craters and anti-tank ditches are impassable to all vehicles and are linear obstacles to infantry. Craters and sections of anti-tank ditch are made crossable by a tipper truck, bulldozer, tank dozer or bridging tank in contact for one move (see **rule 12.27**).

12.8 Permanent obstacles are impassable to vehicles and treated as linear obstacles by infantry. A 4" wide gap may be cleared by a bulldozer or tank dozer in one move or:

- An engineer figure or specialist AFV (e.g. a Panzer I demolition tank) with a demolition charge in contact for one move. Throw a D6. The charge destroys the obstacle(s) for any score but 1 or 2.
- A petard mortar (as mounted in the British Churchill AVRE). Maximum range is 12". Test as for demolition charges to destroy.
- Remote-controlled demolition vehicles (e.g. German 'Goliath' or BV IV). E class with a 6" move, they destroy the target automatically if within 1", but will stop if the controlling two figure team (Goliath) or tank (BV) is not within 18", or is destroyed.

Mount permanent obstacles, such as concrete 'dragon's teeth', onto 4" long strips of thin plywood and landscape their bases.

Playing Tip

Fortifications

These include infantry field positions, bunkers / gun casemates and gun cupolas / turrets.

12.9 Infantry and support weapon elements and anti-tank gun and artillery crews may dig themselves hard cover slit trenches and weapon positions in two moves. Two engineers take three moves to dig in a tank or SP gun.

12.10 Individual concrete or timber and earth bunkers / casemates are always pre-constructed. They may be destroyed by engineers in the same ways as permanent obstacles (see **rule 12.8**), for the loss of all occupants. They are also vulnerable to certain weapons (see **rules 18.16 and 18.17**).

12.11 Armoured gun cupolas may be destroyed by an engineer figure with a hollow-charge demolition device in contact for one move. Test as for demolition charges to destroy (see **rule 12.8**). Gun turrets on bunker / emplacements are destroyed with them. If separate, (e.g. emplaced Panther turrets) they may be fired on as partially concealed tank targets by AT weapons (see **rule 8.44**).

Top: a Crab flail tank begins clearing a marked minefield. It cannot fire whilst flailing.
Inset: The alternative method. A British engineer mine detector in action. A two figure team allows for casualties.

Minefields

12.12 Mines are used to stop, slow or channel enemy movement. They are simplified into one 'dangerous to all' category.

12.13 A single 4" square marked minefield, or an unmarked mined area can be laid or prepared by a minimum of two engineers in two moves. Dummy minefields may also be placed on the table.

12.14 All elements cross a marked or dummy minefield at a maximum speed of 6" per move. All movement should be completed before any casualty tests are taken.

12.15 An unmarked mined area is a point decided upon by the player responsible and must be marked on their command map (see **rule 10.1**). When any element passes within 1" of this point the player must tell their opponent they need a casualty test, which is not taken until the element's movement has been completed.

12.16 Immediately after any movement stage in which an element is in a marked minefield, or passes within 1" of an unmarked mined area, throw one D6 per individual figure, horse or vehicle:

TABLE 13: CASUALTIES FROM MINES

D6 Throw	Effect
1 or 2	The infantry figure becomes a casualty.
5 or 6	The soft-skin vehicle is destroyed and D6 crew / passengers are lost.
4	The AFV sustains light damage.
5	The AFV sustains heavy damage*.
6	The AFV is destroyed*.
*	Place vehicles back inside marked minefields and within 1" of unmarked minefields.

12.17 A single 4" square minefield (or an unmarked minefield that has caused one casualty test) may be cleared by a minimum of two engineers, four infantrymen, or one engineer with a mine detector. Throw a D6 for the number of moves to clear.

12.18 A mine-clearing flail (Sherman Crab) or mine-roller tank (Soviet PT-34) can clear a tank-wide lane through a single marked minefield, or clear an unmarked mined area (that has caused one casualty test) in one move. They may not move into contact with a minefield and start clearing it in the same move. Flail tanks may not fire when mine-clearing.

4" Barbed wire options.

Bridge Demolition and Repair

12.19 Preparing a stone / concrete bridge for demolition takes a minimum of two engineers (with vehicle) three moves in contact (steel bridges take four moves and timber / pontoon bridges two). The charge may be detonated in any move after this and demolishes the bridge for a D6 throw of two or more. If this fails, a second test may be taken in the next move only. The demolition charges may be not be observed by enemy units at any range.

12.20 If a bridge is weakened by HE fire (see **rule 8.90**), it may be made passable for all vehicles by a minimum of four engineers (with vehicle) in three moves.

Bridge Building

12.21 The many types and capacities of military bridges are simplified into two types that can be used by figures and vehicles, including all AFVs. These are:

- Pontoon bridges (British Bailey, US Treadaway, German Brukengerat etc) capable of bridging rivers.
- Light bridges carried by tanks or constructed by engineers, capable of bridging craters, anti-tank ditches and streams.

Pontoon Bridges

German Bridging Column

Heavy truck carrying bridging parts towing a trailer carrying a pontoon section.

+

Half-track carrying engineers towing a trailer carrying a pontoon section.

12.22 A pontoon bridging column capable of bridging rivers up to 12" wide consists of enough horse drawn or motorised vehicles (with or without trailers) to transport two model pontoon sections, one vehicle load of bridging parts and eight engineer figures. Like the bridging column, the model pontoon sections are only representative.

12.23 Bridge building may not take place until a friendly bridgehead has been established on the enemy side of the river. The forward line must be at least 6" from the river and held by a minimum of eight figures. If casualties reduce the defenders to less than eight, bridge building must halt until they are reinforced.

12.24 To start construction, debussed engineers and pontoon-carrying vehicles must be within 3" of the 'friendly' river bank at the point where the bridge is to be built. A minimum of four engineers, or two engineers and four others are required to carry out the construction. Bridge building starts in the following move. To bridge a river up to 6" wide takes four full moves:

Move 1:	One pontoon section is unloaded. Place two pontoons on the river, side by side (1" apart), near the friendly bank.
Move 2:	A 4" length of bridge is fitted across the pontoons.
Move 3:	The pontoons and bridge are floated into position.
Move 4:	Connect them to the banks and complete ramps.
Move 5:	Bridge ready for vehicles to cross at cross-country speed.

12.25 Moves 1 to 4 are classed as construction time. To bridge a river between 6" and 12" wide, double the construction time (ready on move 9). To bridge a river between 12" and 18", triple the construction time (ready on move 13). Two bridging columns are required to build a bridge over 12" long.

Timber Bridges and Tank Bridges

12.26 Timber bridges are constructed in three moves by a minimum of three engineers with their equipment-carrying vehicle within 3" of the obstacle. They can be built across gaps as wide as 4", including small rivers, streams and anti-tank ditches.

12.27 Tank bridges (such as the British Churchill AVRE SBG) may be launched in one move if the tank is within 1" of the obstacle. They can be used to bridge obstacles up to 4" wide.

Pontoon Ferries

12.28 A bridging column carries enough equipment to build two pontoon ferries (instead of, but not as well as a bridge).

- A ferry consists of two pontoons carrying a 4"-5" section of bridge.
- To construct (and operate) a pontoon ferry takes a minimum of two engineers, or one engineer and two others, three moves (as for bridge building, but ready on move 4). Two ferries may be built simultaneously if sufficient figures are available.
- A ferry may carry a tank, an artillery piece and prime mover, or two light trucks / cars.
- Ferries cross at 3" per move and loading and unloading each take one move.

Targeting Pontoon Bridges and Ferries

12.29 Because of the relative size of pontoon bridges and ferries, the target for all HE weapons is always the central 4" section. HE fire or bombing that normally uses the fire grid must instead throw a 5 on a D10 to achieve a hit.

12.30 Two or more HE or bomb hits during a move delay construction of a bridge or ferry by one move, or put a completed bridge or ferry out of action for one move. Working ferries are moving targets. Repairs take one move with a minimum of two engineers.

12.31 Engineers or infantry constructing bridges or ferries are counted as in soft cover. Any elements in the central 4" test for casualties (as normal) after every HE or bomb hit. Figures elsewhere on the bridge may be targeted by small arms and MGs. All elements on the banks may be targeted by all weapons as normal.

British Bridging Column

Pontoon lorry carrying two pontoon sections.

+

3 ton lorry carrying bridging parts.

+

3 ton lorry carrying engineers.

Pontoon Bridge Building

Top left: Troops cross the river to form a bridgehead as the bridging column moves to within 3" of the riverbank.
The next four photographs show the construction moves.
Bottom right: the bridge ready for use.

Assault Boats

12.32 Assault boats are used by infantry to cross rivers. They include rigid, folding and inflatable boats. Some assault boats are powered.

- Medium trucks can carry four boats and tow four more (or two powered boats) on a trailer. They take one move to unload and four figures can move a boat 3" in a move.
- Boats can carry six figures (support weapons equal two figures). Small inflatable boats carry four. To move on water, boats need a minimum of two figures and powered boats a crew of one figure.
- It takes one move to load a boat. Infantry disembarking may move 3" and fire (at short range only).
- Boats may move up to 3" and powered boats up to 6" (12" for the German Sturmboot) per move.
- Infantry in boats are in soft cover. If boats are hit by HE two casualties means all figures and equipment on board are lost
- Two boats (but not small inflatables) may be joined by light bridging to carry a small vehicle or a light infantry or an AT gun. Loading / unloading takes one move. Maximum speed on water is 3" (6" if powered).

Improvised River Crossing

12.33 To improvise a river crossing an element must spend one full move within 3" of the river bank to prepare barrels, rafts etc. They then cross at 2" per move, classed as a target in the open. Non Russian or Japanese units need 5 or 6 on a D6 throw to cross in this way.

A Crocodile flamethrower attacks a casemate supported by various ARVEs (Armoured Vehicle Royal Engineers).

German LMGs

13.1 The German MG34 and 42 had a very high rate of fire and were more effective than other nations' LMGs. To simulate this, add 1 to the fire effect D6 throws for German infantry and 2 for German Panzer Grenadier small arms firing groups (Panzer Grenadiers had two LMGs per squad).

Submachine Gun and Assault Rifle-Armed Units

13.2 If historically accurate, individual companies may be designated as being mainly armed with SMGs or assault rifles (German units from late 1944 only). These companies have the following small arms firing group values:

SMG groups:	+1 to short range values and no long range fire.
Assault rifle groups:	+1 to short range values.
Additions for 'elite' and deductions for 'poor' units still apply.	

Snipers

13.3 Each regiment (or British brigade) may have one sniper figure.

134 Snipers' positions are marked on the command map. They may only be observed after opening fire, at a range of up to 24", for a throw of 5 or 6 on a D6. This test may be repeated every move the sniper stays in that position. Once observed, snipers can only regain concealment if they can move under cover (i.e. within buildings or woods) to a new position.

Petrol Bombs

Known as 'Molotov cocktails' after the Soviet politician who promoted their use, petrol bombs were most used by Soviet troops up to 1942, home defence forces (like the Home Guard and Volksturm) and partisans.

13.5 Snipers must indicate a specific, single figure target before firing, which may include COs. To hit a target throw a D6. The score needed is 4, 5 or 6 for a figure in the open, 5 or 6 if in soft cover and 6 if in hard cover. Snipers are not included in any morale test.

Petrol Bombs

13.6 Petrol bombs can only be used by defending units and replace the grenade 'allowance', with all the same limitations (see **rules 8.21 to 23**). A battalion using petrol bombs - even if only in one company - may not use AT grenades.

13.7 Petrol bomb range is 3". They may be thrown at any moving or stationary vehicle and into buildings (not bunkers). Throw a D6 to decide combined hit and effect. **Table 14** is used to determines casualties and damage.

TABLE 14: PETROL BOMBS

Target ↓ D6 Throw →	1, 2, 3 or 4	5	6
Figures	No effect	1 casualty	2 casualties
Soft-skin vehicles		Destroyed	
Armoured vehicles		Light damage	Destroyed
Buildings (not bunkers)		On fire	On fire

This section introduces new forms of fire support, restrictions on different armies' use of artillery and additional rules for IDF. It will be particularly useful for larger games. Under the basic artillery rules, units that form part of a Rapid Fire force provide Direct Support Fire (DSF) to its elements. However, many armies in WWII also deployed artillery units that could provide General Support Fire (GSF).

14.1 GSF is additional pre-arranged or requested fire from 'shared' artillery units at divisional or corps level. If permitted, an OP may request one move of GSF by throwing a D6 and testing (see **Table 15**).

- Agree how many GSF guns (+ calibre) are available before the game begins.
- Any prearranged GSF must form part of a Fire Plan (see **rule 14.10**).
- An OP declares the number and calibre of available GSF guns it is requesting.
- If successful, the fire will be available for the OP to use in their next turn without the need for a radio test.
- A maximum of three moves of requested GSF may be used in a game.

TABLE 15: OP REQUESTED GSF

Army	Maximum DSF Guns Per OP Per Turn	Maximum GSF Guns Per OP Per Turn	Maximum GSF OP Moves	D6 Throw To Request GSF
British / Commonwealth / US / Late war Allies*	12 *	8 *	3	4, 5 or 6
German	3	2	2	5 or 6
Other Axis / 1940 French / Belgian / Finnish / 1939 Polish	3	1	1	6
Russian***	5	-	-	-
Japanese	3	-	-	-

* From 1943. Before then British / Commonwealth OPs may only control up to four DSF guns and / or two GSF guns and American OPs up to six DSF guns and / or three GSF guns at a time.

** If more than one OP requests GSF the first to successfully test gets priority for the available guns.

*** IFrom 1942, Russian forces may use GSF guns and heavy mortars for predicted and pre-registered fire only. In the attack, after any fire plan has been completed (see **rule 14.10**), Russian units will use direct fire weapons to engage enemy elements. In 1945 a Russian force may add one tank OP with radio.

If desired, one or more GSF artillery units can be positioned on table if there is enough space behind the 'frontline' forces for them to deploy. Place medium / heavy guns behind field / light units.

Playing Tip

Predicted Fire

14.2 Predicted or 'map' fire is a type of IDF used when no friendly OP can see the target. In real life it relies on complex calculations and is less accurate than observed fire.

- Up to five moves (per battery / battalion) of predicted fire may be used in a game, but not by integral infantry battalion or regiment guns. On-table guns that are moved may not be used for predicted fire, unless British/Commonwealth/US and static for five moves (from 1943 only).
- Predicted fire may be carried out by the DSF guns of the on-table force (i.e. a British artillery regiment supporting a brigade) and by GSF guns. It is particularly used for pre-game and pre-assault bombardments.

14.3 Targets may be:

- A geographical feature, or identifiable part of a feature (i.e. a building, farm, wood, road junction, bridge, corner of a hedge, northern end of a hill etc).
- Fixed enemy positions, such as gun or infantry emplacements and bunkers.

> Predicted fire may be used against off -table guns. Set the guns out on a seperate flat surface and fol- low rules **rules 14.4 to 14.6.**
>
> Playing Tip

14.4 No test is required to target geographical features, but in the case of enemy positions a test must be taken for each potential target. Agree the level of the firer's intelligence information. Throw a D6. The following scores are needed to fire at the target:

Intelligence	Score
Good intelligence (e.g. for the Allies in Normandy).	4, 5, or 6
Average intelligence (e.g. for the Allies in Op Market Garden).	5, 6
Poor intelligence (e.g. for the Germans in the Battle of the Bulge).	6

14.5 To fire at the targeted feature or enemy position use the plastic fire grid. Lay the middle square (5) over the centre of the identified target and throw a D10 for each gun firing to decide in which grid square its shot lands. 0 is a miss.

14.6 If an enemy element is hit, select the appropriate table on the casualty chart and throw a D6 to determine the effect on the target. (If there are more than one possible targets in the square throw a D6 to decide which is hit: e.g. 1/2 3/4 5/6 for three potential targets). The modifiers below give possible deductions from the score.

Modifiers for Predicted Fire	
Deduct 1	from the throw for any units hit by fire against a geographical feature.
Deduct 1	from the throw if a unit is hit that is not occupying the enemy position being targeted.
Deduct 2	from the throw for any hits on moving units.
The usual deductions for cover still apply.	

Barrages

14.7 The barrage is a curtain of predicted fire (and / or smoke) that advances ahead of an attacking force. These barrages can be no longer than five moves and must be planned before the game begins by:

- Marking the barrage start line and the barrage's direction on the command map (see **rule 10.1**).
- Writing down the moves on which the barrage will start and end and how far it will move forward each turn.
- Noting the total number and calibre of the guns available to fire and how many of each calibre will fire into each 12" square (see example on page 92).
- Identifying which squares (if any) will have smoke fired into them and on which moves. (For smoke, use barrage firing procedure during the smoke stage of each turn)

Firing a barrage:

- Divide the area along the enemy side of the barrage line into 12" x 12" sections, laying the plastic fire grid over each in turn.
- Throw a D10 for every gun firing into each 12" grid square. Each throw will decide where a round lands. 0 = miss. Test for effect on any element that is hit, with normal IDF deductions (if more than one target see **rule 14.6**).
- If a target falls under the junction of several squares, decide which square will count as a hit before throwing the D6.
- For a moving barrage, move this line forward the planned distance at the beginning of the next turn's firing stage (or smoke stage, if smoke is being used). Repeat the firing procedure for the specified number of moves.
- Once started, a barrage may be stopped (see **Table 16**) but not modified.

Army	British / US	German / Italian / Axis / French / Japanese	Russian
Moves to stop	Next turn	1	2

TABLE 16: CANCELLING A BARRAGE

A BARRAGE

A British three-move moving barrage fired by twelve guns advancing 12" per move

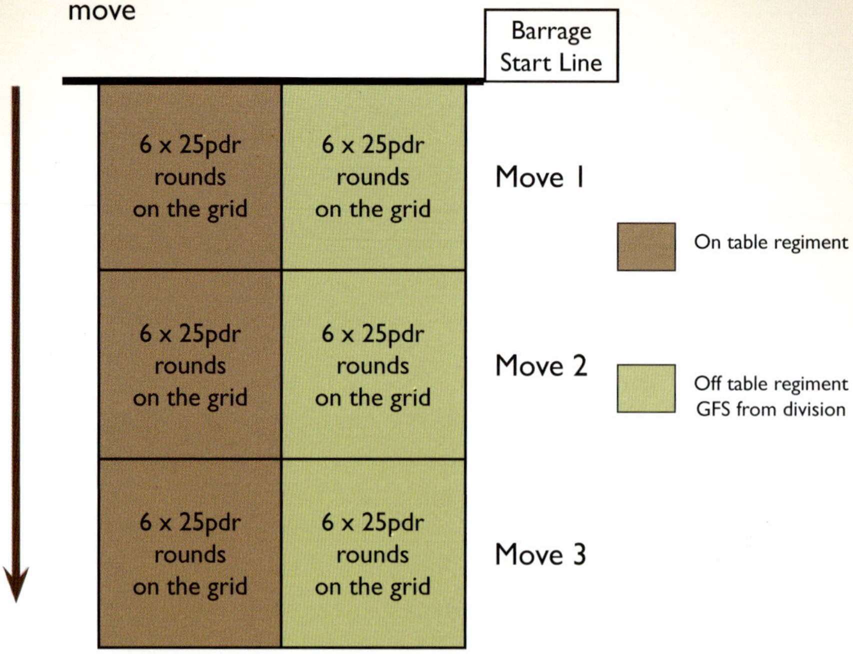

HE explosion markers (made from hamster bedding!) remind players that firing is taking place for both fire effect and morale purposes.

Pre-Registered Fire

14.8 This may only be used by OPs that are occupying established positions at the beginning of the game, or (from 1943) by British, Commonwealth, US or other late war Allied OPs after five moves in one position. Each OP may register five points of aim it can observe. Mark them on the Command Map.

- Points should be easily recognisable features (i.e. the middle of a road junction, the corner of a wall) or the centre of an observed enemy position.
- Any unit (moving or static) observed within 2" of a point may be a target.
- Throw a D6 per 'shot'. 4, 5 or 6 (5 or 6 if target moved last turn) means a hit. Test for effect as normal.
- GSF guns may also be used (see **Table 15**), but only if in a fire plan unless British, Commonwealth, US and late war Allies (from 1943).

Use the risk zone circle, centred on the point of aim, to check what may be a target for pre-registered fire.

Playing Tip

American OPs

14.9 To represent the US move to train infantry as artillery observers, US battalion HQs may act as OPs from 1943 onwards. The radio test requires a 1 or 2 on a D6 to succeed. Only a single DSF artillery unit may be controlled each turn.

Fire Plans

14.10 To support an assault against a prepared defensive position the attacker may use a combination of artillery support. A fire plan is a timetable that lists what types of fire are to be used and when.

Example of a Fire Plan

A British Brigade is to attack a German defensive position that has been constructed in front of and behind a ridge that prevents observation. It will attack with two battalions supported by a squadron of tanks and a field regiment of 25pdrs. A medium battery of 5.5" guns is available for General Support Fire. Three OPs are in action: one with each battalion and one with the tanks.

Move 1.	*Pre-registered fire against 1 enemy position per OP. Each OP controls a battery of 2 DSF 25pdrs and a battery of 2 GFS 25 pdrs.*
Moves 2-4	*A 3-move, 24" wide predicted fire moving barrage advancing 6" per move (infantry speed) on the enemy forward positions and over the ridge. Fired by: 2 x 25pdr field regiments (12 guns) and 5.5" medium battery (4 guns)*
Move 5	*Predicted fire against road junctions and identified enemy gun positions on the far side of the ridge. Fired by all guns.*
Move 6	*Observed IDF, from the OPs (now in new positions on top of the ridge), against enemy positions and units on the far side. This may include OPs calling up other batteries (and requested GSF from the next turn).*

The GSF guns are prearranged, so don't need to be requested. (→ Move 1)

Each 25pdr regiment fires a 12" wide lane of the barrage. The 5.5" guns can be allocated two per lane to thicken the fire. (→ Moves 2-4)

All guns are eligible for predicted fire as they haven't moved. Only the enemy positions will have to be tested for. (→ Move 5)

The OPs moved up during the barrage and predicted fire. They now use Basic Rules IDF with their DSF guns, plus any requested GSF support. (→ Move 6)

A British 5.5" battery deployed on-table for GSF.

Use of Aircraft

15.1 Each aircraft model represents six real machines, except for observation planes which represent one. Aircraft carrying out attacks or observing must be on-table

15.2 Each side is allowed a maximum of one combat aircraft and one AOP over the table per move. However, for large games this may be increased by mutual agreement.

15.3 Aircraft arrive over the table at the beginning of a turn, either:

a) On a move planned before the game (see **rules 10.3 and 15.9**), or

b) As the result of a request for air support via an 'air link' vehicle that can observe the selected target(s) (see **rule 15.12**).

15.4 In the case of b) the aircraft will not arrive until four moves after being requested (i.e. requested during move 3, arrives on move 7). In both cases, on the move the aircraft is due to arrive test for delays. Throw a D6. 4, 5 or 6 means it arrives one move later.

15.5 Aircraft may operate over the table for a maximum of five moves. These include any moves in which the aircraft is engaged in air-to-air combat or is driven off for a move by air-to-air combat or AA fire (see **rule 15.27**).

15.6 The altitude of an aircraft is determined by its role during a move. In game terms aircraft operate at either low or medium altitude.

Model aircraft can be mounted on thin, clear plastic 'poles' fixed to moveable bases. The height of these poles doesn't represent the altitude of the aircraft

Playing Tip

Ground Attack

15.7 Air attacks take place during the firing stage of a turn. Ground targets for aircraft are selected by the player controlling them. Target types are the same as those for HE fire (see **rule 8.3**) and must be identified immediately before firing or bombing begins. No measuring is required.

15.8 To identify the selected target throw a D6. The target is spotted by the aircraft for anything but 1. If it fails it will climb out of AA range and circle the area (but will still engage in air-to-air combat). It may test to spot in the next move. If the target has been marked by a coloured artillery smoke shell (1943 onwards) spotting is automatic.

15.9 If pre-ordered, the aircraft may only attack a maximum of three targets in total, descriptions of which must be written down on the move before the plane is due to arrive on table. For example:

Target 1 - Infantry position north of village A

Target 2 - Tanks north of hill B

Target 3 - Gun emplacement west of wood C

15.10 The sequence of play for ground attack is as follows:

1. Identify target and test for spotting.
2. Place the aircraft's base within 6" of the target.
3. The enemy carries out all reserved anti-aircraft fire.
4. If not driven off or shot down, the aircraft carries out its attack.
5. The aircraft remains in its attack position to receive any AA fire during the enemy's next turn.

15.11 Aircraft may not attack the same target on consecutive turns, but may return later (i.e. a plane may attack the same target in move 1, move 3 and move 5).

15.12 If requested by an 'air link' OP vehicle the aircraft may attack any target observed by the crew of the vehicle. A radio test (see **rule 8.68**) must be successfully taken and the target clearly identified before any attacks are carried out. If the 'air link' vehicle is destroyed the aircraft must test to acquire a new target of the player's choice each move (a throw of 4, 5 or 6 on a D6). No 'spotting' throw is required.

15.13 In a single move an aircraft may carry out either a strafing, anti-tank / rocket, or bombing attack.

Strafing (Low Altitude)

15.14 Aircraft armed with machine guns or cannon (usually fighters) carry out strafing attacks on ground targets at medium range. Hard / soft cover applies. See **Table 17** for points values of guns.

TABLE 17: STRAFING

Armament	Points Value For Small Arms Firing Effect
2 forward-firing MMG	8
2 forward-firing HMG	10
3 - 4 forward-firing MMG	10
3 - 4 forward-firing HMG	12
More than 4 forward-firing MMG	10 (two throws)
More than 4 forward-firing HMG	12 (two throws)
Single forward-firing autocannon	8
Multiple forward firing autocannon	15

Anti-Tank (Low Altitude)

15.15 Some aircraft are armed with anti-tank guns or rockets. **Table 18** gives the minimum D6 throws to achieve an effective hit against an AFV. After achieving an effective hit throw another D6 to determine damage as per normal anti-tank firing. No AT firing modifiers apply.

Table 18: Aircraft Anti-Tank Guns and Rockets

AT Armament ↓ Armour Class →	A	B	C	D	E	F (soft-skin)
Multiple HMGs / single 20mm autocannon					6	5
Multiple 20mm autocannon			6	5	4	3
Single or twin 30 – 50mm gun(s)		6	5	4	3	3
75mm gun(s)	6	5	4	3	3	3
Rockets*	6	5	4	4	4	4

* Aircraft carry eight rockets. Must be fired in multiples of two. Throw one D6 per rocket.

15.16 Rockets may also be fired against other targets and have a HE effect of 10 points on table 4. To achieve a hit (hits) roll two D6. 4, 5 or 6 = a hit.

Dive Bombing (Low Altitude)

15.17 Dive bombing is the most accurate method of bombing and is carried out by aircraft diving on a target at anything from 30 degrees to near vertical. Dive bombers may only drop one per turn. For determining hits throw a D6 for each bomb. 4, 5 or 6 = a hit. The skip bombing method (only used from 1944 by superior fighter bombers armed with a single bomb) is treated in exactly the same way.

Level Bombing (Medium Altitude)

15.18 Level bombing must be pre-ordered and can be carried out by bombers and fighter bombers. As the model plane and its bomb load is representative, any number of bombs may be dropped in a move, from one to the maximum number carried by the aircraft.

15.19 To drop bombs, place the plastic fire grid so the centre square (5) is over the target and the grid is aligned with the direction of attack. Throw a D10 to decide where the bomb(s) land(s).

15.20 Bombs are divided into five types: small, light, medium, heavy and anti-tank. **Table 19** below gives their points value and which table of the casualty chart to consult when calculating effect. Throw a D6 to determine casualties / effects on soft-skin vehicles. If a gun or support weapon is hit throw another D6. 6 = destroyed.

Table 19: Bombing

<table>
<tr><th>Bomb</th><th>HE Effect*</th><th>Anti-Tank Effect (D6 Throw)</th></tr>
<tr><td>Small</td><td>8 points table 4</td><td>No effect</td></tr>
<tr><td>Light</td><td>10 points table 4</td><td rowspan="3">1 or 2 = light damage
3 or 4 = heavy damage
5 or 6 = destroyed</td></tr>
<tr><td>Medium</td><td>12 points table 5</td></tr>
<tr><td>Heavy</td><td>15 points table 5</td></tr>
<tr><td>Anti-tank (PTAB)**</td><td>No effect</td><td>1 = light damage
2 or 3 = heavy damage
4, 5 or 6 = destroyed</td></tr>
</table>

* Deduct 1 from the D6 casualty throw if the target is in hard cover

** PTAB bombs were a Soviet weapon first used in July 1943.

15.21 Aircraft bomb loads varied greatly. For the purposes of the game use the following maximum loads:

Bomber Type	Bomb Load - Option 1	Bomb Load - Option 2
Dive bomber	2 light, 1 medium	1 heavy
Fighter bomber	4 small	2 light
Superior fighter bomber	2 medium	1 heavy*
Light bomber / ground attack	4 light	2 medium
Medium bomber	6 medium	3 heavy

* Only permitted load for low level, 'pebble skimming'–style skip bombing attacks.

Examples of 1944 Aircraft

British Typhoon
Type: superior fighter bomber
Armament: 4 x 20mm autocannon and rockets
(may carry bombs in place of rockets)

American Thunderbolt
Type: superior fighter bomber
Armament: 6 x HMG and rockets
(may carry bombs in place of rockets)

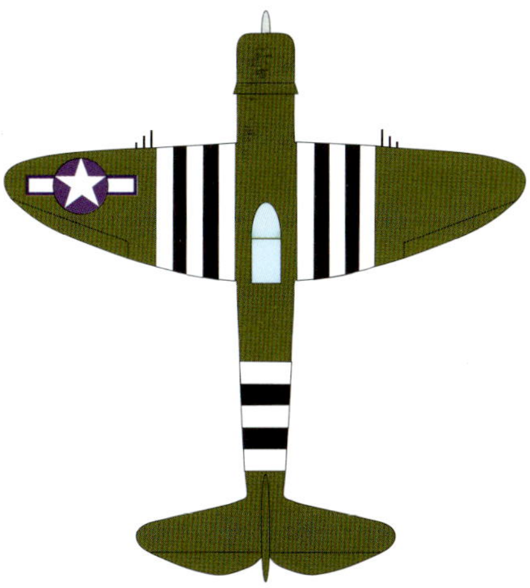

German FW 190G-2
Type: superior fighter bomber
Armament: 2 x 20mm autocannon
2 x HMG and 1 medium bomb

American Boston (in British Service)
Type: medium bomber
Armament: 4 medium or 2 heavy bombs

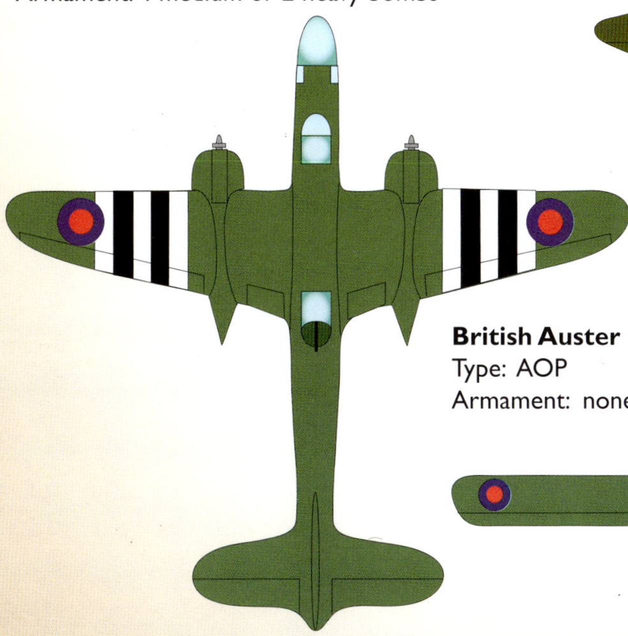

Russian Shturmoviki
Type: ground attack
Armament: 2 x 37mm and rockets
(or 2 x PTAB bombs in place of rockets)

British Auster
Type: AOP
Armament: none

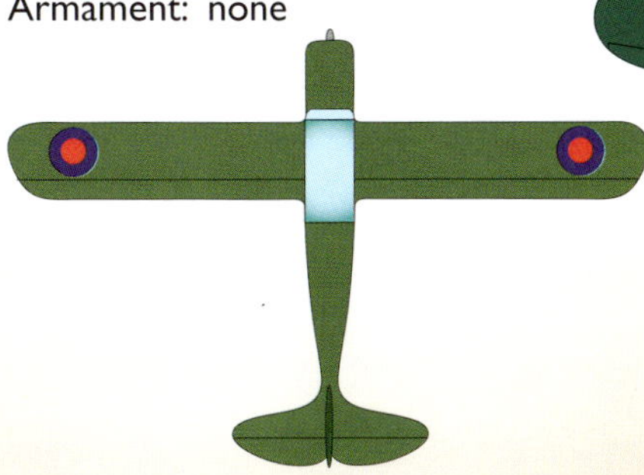

Air Observation Planes (Medium Altitude)

15.22 Aircraft must be unarmed specialist air observation planes (AOPs) such as the German Feisler Storch, British Auster and US Piper L4. AOPs may be positioned anywhere on the table. Test to spot targets during the observation stage of a turn.

15.23 An AOP may be used to observe any enemy units on the table and inform either an artillery relay or 'airlink' vehicle of their position. Throw a D6 for radio contact each move. 1 = failure. An AOP may observe for artillery (see **rule 8.66**) or ground attack aircraft during its five moves on table: not both.

15.24 Aircraft may automatically observe any unit that is moving or firing, except for infantry units in woods and buildings (which, even if firing, cannot be observed). Otherwise, a test must be taken to observe successfully. To carry out an air observation test throw a D6 and consult **Table 20** below.

TABLE 20: OBSERVATION FROM AOPs

Target	D6 Throw
Static vehicles / guns / mortars in the open	2, 3, 4, 5 or 6
Static infantry units in the open	3, 4, 5 or 6
Static vehicles / guns / mortars amongst buildings	4, 5 or 6
Static camouflaged vehicles / guns / mortars	5 or 6
Infantry dug in or amongst buildings	6*

* 5 or 6 if firing.

The anti-aircraft guns do not include large calibre dual purpose weapons such as the German 88. These were generally used to engage high altitude targets rather than fighter bombers and if any are 'on table' they should be occupied with ground targets

Design Note

Anti-Aircraft Fire

15.25 Ground fire against an aircraft may be carried out by:

- Any deployed infantry or artillery unit - not in vehicles, buildings or being towed - that it is about to attack.
- A dedicated anti-aircraft gun (HMG on AA mount and larger) within 36" of a unit being attacked by an arcraft at low altitude.
- A dedicated, single-barrel, 37mm – 40mm AA gun which tests to target an AOP or bomber at medium altitude (see **Table 21**).

15.26 Aircraft can either be driven off (table) for one full move by AA fire or shot down. An aircraft driven off by reserved fire (see **rule 6.1**) may not continue its attack that turn. AOPs driven off may not observe next turn.

Anti-Aircraft Fire		D10 Throw Required	
Unit / weapon	Target altitude	To drive off	To shoot down
Infantry unit	Low only	7 or 8	9 then 4, 5 or 6 on a D6
Artillery unit	Low only	7 or 8	9 then 4, 5 or 6 on a D6
Single AA HMG / 20 - 30mm autocannon	Low only	7 or 8	9
Multiple AA HMG / 20 - 30mm autocannon	Low only	4, 5, 6 or 7	8 or 9
Single-barrel 37 - 40mm AA	Low or medium	7 or 8	9

- Dive bombers cannot be driven off – only shot down.
- Throw a D6 to test if a single-barrel AA gun has the range to target an enemy AOP. 4, 5 or 6 means it may fire.
- The Soviet IL-2 Shturmovik and German Hs 129 ground attack aircraft were heavily armoured. Any successful D10 fire effect throws are reduced by 1 (e.g. a throw of 3 becomes 2, 2 becomes 1 and a 1 means no effect).

TABLE 21: ANTI-AIRCRAFT FIRING

Most air-to-air combat took place well away from the battlefield, trying to stop attack aircraft reaching their targets. See **rule 19.13** for a simple way of simulating the effects of this more distant combat.

Playing Tip

An American M16 SP multiple HMG AA gun covers US troop movements. France 1944.

Over the Mediterranean (for a change), a Sea Hurricane fighter drives a JU 88 medium bomber off-table.

Air-to-Air Combat

15.27 Air-to-air combat occurs if two opposing armed aircraft are on table at the same time. Both sides simultaneously test for combat every turn until a conclusion is reached. Aircraft in an air-to-air combat may not attack ground targets and vice-versa. Bombers may not fight bombers.

15.28 To fight an air-to-air combat both players throw a D6, modified according to aircraft type (see **Table 22**). To drive an enemy plane off-table (so it may not return for one full move) a player must get a score that is two clear points higher than his opponent's throw (e.g. a score of 5 versus a score of 2, not 5 versus 3).

Table 22: Air-to-Air Combat

Aircraft Type	D6 Modifier
Superior Fighter	+1
Fighter	0
Light bomber	-1
Medium / heavy bomber	-2

e.g. Fighter A can drive fighter B off table if A throws a 4 and B throws a 1 (two clear points). Light bomber X can drive off attacking superior fighter Y if it throws a 6 (modified to 5) and the superior fighter throws 1 (modified to 2).

- Dive bombers are treated as bombers. Skip bombers are superior fighters.
- Superior Fighters are the fastest and / or better armed fighter aircraft.
- If an aircraft is driven off for a second time it is considered to have been shot down.
- Fighter bombers that ditch or have previously dropped their bomb(s) count as fighters. Otherwise treat as light bombers. Ditched bombs have no effect on ground targets.
- Observation aircraft which are their side's only aircraft on table are automatically and permanently driven off-table by all fighters. However, if a friendly combat aircraft is also on table the enemy aircraft will fight that first and the AOP can continue.

These rules are designed for games that feature sustained combat in towns and industrial areas.

Mouseholing

16.1 Troops may move between linked buildings without leaving cover by blowing holes in their shared walls ('mouseholing'). This is available to engineers and infantry equipped with pole charges (maximum of two per company, declared before the game). One charge in position for one move will blow a hole in a wall between two buildings (no test is necessary) but this may not be followed by a close assault move.

British engineers with pole charges prepare to make a mess of someone's living room wall.

Large Buildings

16.2 In a large building with several floors (i.e. apartment blocks), or with sections separated by major internal walls (i.e. factories) survivors of a unit that loses a close assault combat may move to another floor or section. In this case the attacker must carry out another close assault to secure the new floor or section. If the defenders lose again they surrender.

Cellars and Sewers

16.3 Cellars may be represented by ground floor rooms that are only accessible to the floor above. They may have small windows high on their walls and, if declared before the game, exits into a sewer system.

16.4 Cellars offer full protection from HE fire by weapons of 125mm calibre or less and small or light bombs. For larger HE weapons and bombs deduct 2 from the fire effect D6 throw.

16.5 Troops occupying cellars may only fire if there are windows. In close assaults they will surrender if they lose (or withdraw via a sewer exit). If they occupy the floor above, attackers may drop grenades, demolition charges or petrol bombs into cellars, or use flamethrowers. Flamethrowers and other direct fire weapons treat cellar windows as bunker firing slits.

16.6 Sewers systems must be marked on the command map prior to the game, with manhole exits and entrances in streets and cellars clearly marked. Troops travel 4" per move through sewers and a tunnel may be permanently blocked by a successfully detonated demolition charge (see **rule 12.8**).

Airborne troops saw their first combat use in WWII and became a high risk, but potentially highly effective way of launching unexpected attacks against distant objectives.

Airborne Landings

17.1 Any airborne attack must be planned before the game and parachute battalion and glider drop zones (DZs) marked on the attacker's command map.

17.2 No reserved fire is allowed against airborne landings.

Landing Paratroops

17.3 There are two ways in which paratroop landings can be carried out:

17.4 For small scale (1 or 2 battalion) drops prepare 1" square pieces of paper to represent each figure and heavy weapon being landed. Mark on which company or weapon crew the figure belongs to, as well as the CO and any specialists, such as radiomen or engineers.

1. Hold the pieces of paper approximately 24" above your selected drop zone (DZ) and let go. (For paratroops dropped by inexperienced aircrew drop from 36").
2. Replace each piece of paper with an appropriate figure or weapon unless the paper square has landed in water (in which case it is lost), or on an obstacle (building, wood, hedge, wall etc). If the latter, throw a D6. 1, 2, or 3 = figure or weapon lost.

17.5 No paratroop figure may move in the turn in which it is dropped.

17.6 Some early paratroop forces (notably the Germans) dropped personal weapons in containers. Prepare a paper square for each container (one per company) and drop as normal. Troops must reach a container before they can fire small arms.

17.7 For larger scale landings (three or more battalions) an alternative and less time consuming method of landing paratroops is to use the plastic fire grid. In this case the grid represents a battalion drop zone (DZ).

1. Place the grid on your selected spot on the table.
2. Throw a D10 for each company. Scores of 1-9 give the square in which that company lands. Place figures and weapons on the table
3. A test must be taken if the throw is 0, or if any obstacles are within the landing square.
4. Throw a D6. The score = the number of casualties. Throw another D6 for each heavy weapon. 6 = destroyed.

Glider Landings

17.8 To land a glider requires a clear space 6" long (in addition to the model). Place the model glider on the table and throw a D6.

1. A score of 1 means there are casualties.
2. Throw a second D6. The score = the number of casualties.
3. Each piece of heavy equipment, such as an MMG, mortar, jeep, carrier, gun or light tank must be thrown for separately. A D6 throw of 6 = destroyed.

17.9 In the same turn as the glider lands any surviving infantry may move 3" from the glider and engage targets with small arms at short range. Heavy equipment crew figures will take the rest of that turn preparing to unload.

Paratroop and Glider Infantry Morale

17.10 The original strength of parachute and glider units is the number of troops who survive the landing. Figures lost during the drop or glider landing are not counted as casualties for morale purposes.

Para-Dropped Vehicles

17.11 From 1944 onwards Allied SAS units may be equipped with Jeeps dropped by parachute. Select a DZ and use the plastic fire grid (and a D10 throw) to decide where they land. Then throw a D6 for each Jeep. 5 = damaged (two moves to repair). 6 = destroyed. It takes the rest of the dropping move to 'unpack' the Jeep and make it ready to move. Jeep crews drop separately.

A GLIDER LANDING

Left: a Horsa glider carrying an airborne rifle company (eight figures), 6pdr AT gun, Jeep and crew (three figures) lands in an open field. A 1 is thrown and therefore the Horsa has crash landed. A further D6 throw of 3 means three passengers are casualties. Damage throws for the AT gun and Jeep are 2 and 4 respectively so these are undamaged. Five figures exit the glider and move 3".

Right: in the German turn no further casualties have been caused and so during the next British turn the five survivors from the company move their full 6". Having spent the last turn preparing to off-load the equipment the AT crew can now disembark. The Jeep (which is in a prime mover role towing the 6pdr) has a potential maximum movement of 12". As the crew wish to move through the wood a D6 is thrown and the resulting 3 is deducted.

Amphibious Landings are carried out by troops attacking from water to land. They are usually assaults on fortified beaches (as on D-Day and in the Pacific), but can also include landings in estuaries, over lakes, or across very large rivers. In the majority of amphibious landing games only a narrow strip of water (12" – 24") will be represented, as the main focus of the game is on the land assault.

Preliminary Bombardment

18.1 Most amphibious landings were preceded by naval bombardments. Represent them with up to five moves of predicted fire (see **rules 14.2 to14.6**) from six 150mm naval guns, against any identified enemy positions within 36" of the waterline (**rule 14.4** can be ignored).

18.2 Air attacks were also used to 'soften up' enemy defences. These must be pre-ordered with up to three nominated targets (see **rule 15.9**) and may start from move 1. Targets may not be engaged within 18" of the waterline or 12" of the most advanced friendly troops. Before 1944, Allied air and naval gunfire attacks on the same coastal position may not take place in the same turn.

18.3 Landing craft MGs and autocannon may only be used for AA fire.

Landing Craft and Amphibious Vehicles

18.4 Assaulting troops, vehicles and weapons were transported to land in purpose-built landing craft, in amphibious vehicles such as DUKWs and LVTs, or in improvised landing barges and coastal craft. Maximum capacities of the most common landing craft and amphibians used by the Allies are as follows:

Vessel / Vehicle	Maximum Load
Landing Craft Assault (LCA)	16 figures
Landing Craft Vehicle Personnel (LCVP)	16 figures, or 1 Jeep or 1 light gun
Landing Craft Mechanised (LCM)	24 figures, or 1 tank / SP gun / medium truck, or 2 light vehicles* or guns and up to 8 figures
Landing Craft Tank (LCT) Mk 3, 4, 5 or 6	The number of model vehicles or guns that can be fitted into the vessel's tank deck and any vehicle crews
DUKW amphibious truck	12 figures, or 1 gun up to 105mm** and crew
Landing Vehicle Tracked (LVT or 'Buffalo') Mk 2 (no ramp)	12 figures
Landing Vehicle Tracked (LVT or 'Buffalo') Mk 4 (rear ramp)	16 figures, or 1 light vehicle*, or 1 gun up to 105mm and crew

* Light vehicles include Jeeps, 15 cwt trucks US weapons carriers and (for LCMs and LVT Mk 4s only) British carriers.

** DUKWS carrying guns may flounder. Throw a D6 at the beginning of water moves. 1 or 2 = sunk.

18.5 In games with a narrow strip of sea, landing craft and amphibious vehicles take two moves to reach the beach. Troops and vehicles / towed guns may start to disembark during the second move, but may only move up to 4". Infantry cannot fire during this turn.

18.6 Once on land, troops carried in amphibious vehicles may disembark in any move with normal deductions. Troops in LVT 4s with rear ramps deduct 3" for disembarking and count as in soft cover for this move.

18.7 Landing craft and amphibians in the water can only be fired at by enemy coastal guns using IDF. On reaching the beach throw one D6. 1 = the landing craft has hit a mine. See **Table 23** for effects of each IDF HE hit or detonated mine.

Table 23: HE and Mine Damage to Landing Craft

Landing Craft
Throw a D6 for passengers. The score = number of casualties. Throw a D6 for every vehicle / weapon. 6 = destroyed.
DUKW / LVT
Throw a D6. 6 = vehicle and load destroyed.

Left: an LCT 5 with three Sherman tanks enters from the table edge.
Right: On turn two it beaches and the tanks start to disembark and the lead AFV can move 4". See rule 18.7 above for possible damage.

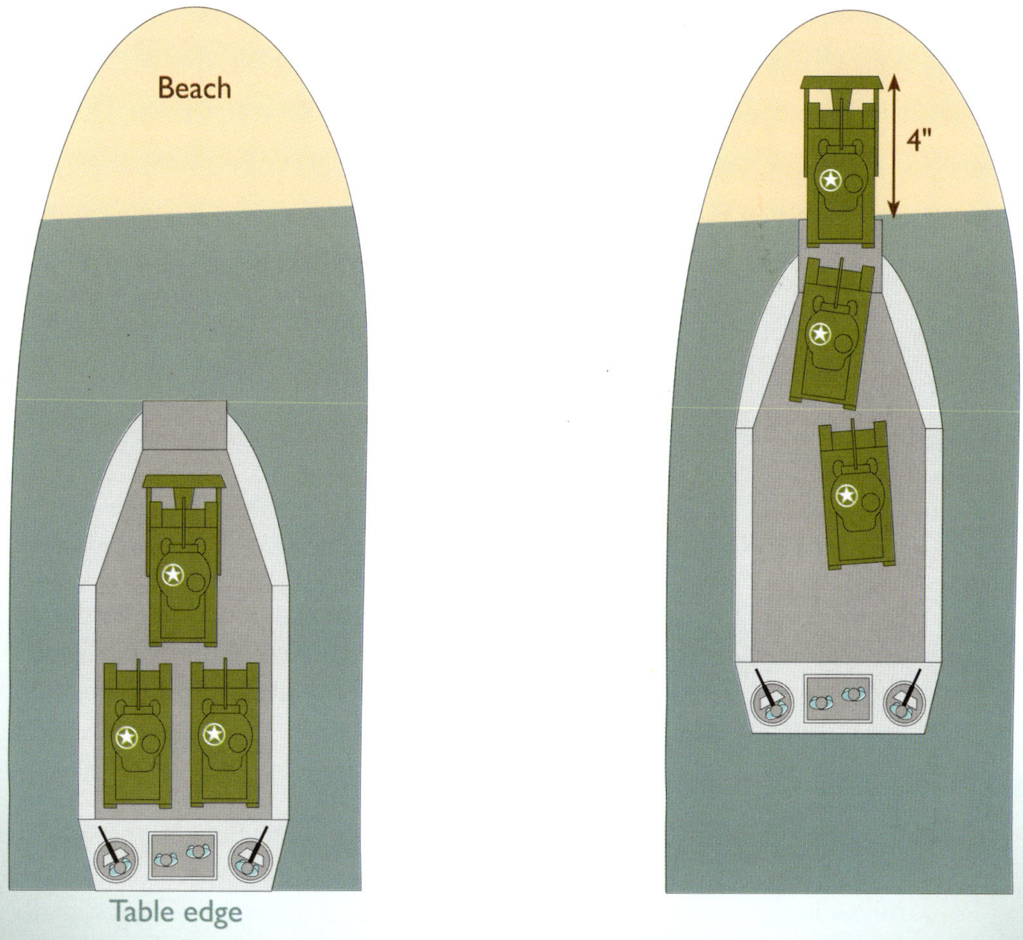

Amphibious Tanks

18.8 The most common of these is the DD (duplex drive) Sherman. It takes two moves to reach the beach and, like the DUKW, can flounder before landing. Whilst the DD is in the water throw a D6 before every movement stage. 1 means the tank sinks. If it survives it may not fire until the 3rd move.

18.9 Armoured, turreted LVT(A)s also take two moves to land, but cannot flounder and may fire their guns at identified targets during the 'run in' to the beach using direct fire (see **rules 8.57 to 8.60**). Deduct 1 from D6 throw for trying to achieve a hit.

On the Beach

18.10 Maximum movement for infantry and vehicles on beaches is 6" per move.

18.11 The rules assume that any defended beach will be mined. Throw a D6 for every moving vehicle (except any flailing Crab tanks) per move.
1 = destroyed, 2 = immobilised for the rest of the game, but may still fire.

18.12 Mines may be cleared by a Crab flail tank or bulldozer at 4" per move, or by a group of four engineers at 2" per move. When the clearing vehicle reaches the far edge of the beach, that sector (agreed and mapped pre-game and up to 30" wide) is declared cleared. Vehicles following a Crab flail tank or moving through a cleared sector need not test. Infantry do not test for mines on beaches

18.13 Beach obstacles, such as girder 'stars', offer no cover and are only representative. Infantry may not dig-in on beaches.

18.14 Anti-vehicle concrete or palm log walls along the rear edge of beaches are impassable to all vehicles, but can provide soft cover to infantry. A vehicle wide gap may be blown by two engineers working for two full moves to prepare a demolition charge. Test as for obstacle demolition (see **rule 12.8**). Alternatively, the wall may be crossed by deploying a tank bridge or ARK

Bunkers and Emplacements

18.15 Concrete, earth and palm log bunkers / gun casemates and Tobruk MG positions are bullet proof (small arms fire cannot cause casualties).

18.16 Bunkers and gun casemates can be destroyed by demolition charges and petard mortars (see **rule 12.8**), or by naval gunfire. If hit by a naval gun, throw a D6. 6 = destroyed. Destruction of a bunker or casemate means the automatic loss of all occupants.

18.17 Guns of 75mm (or greater) calibre, infantry anti-tank launchers and flamethrowers within range can attempt to fire into bunker and casemate embrasures and a group of not less than four figures within 2" can try to throw in a grenade. In all cases throw a D6. 6 means all the occupants are lost and all weapon(s) are destroyed (4, 5 or 6 for a flame thrower).

18.18 When firing HE against troops occupying trenches and open emplacements deduct 2 from the D6 throw for effect (rather than the usual deduction of 1 for hard cover).

Morale

18.19 Immobilised, or heavily damaged vehicles on beaches are abandoned by their crews for a D6 throw of 1 or 2. Throw a D6 in every subsequent turn. 5 or 6 means they return.

18.20 Normal infantry morale rules do not apply whilst on the beach. Test for infantry morale by individual landing craft 'load' as set out in **Table 24**.

Table 24: Beach Morale

D6 Throw Required...	Score	Effect
After four casualties	1 or 2	No advance for one turn and abandon one support weapon
After eight casualties (and each following turn)	1, 2 or 3	Boat load pinned down and abandon all support weapons

- Other scores have no effect on the attackers.
- Abandoned weapons cannot be recovered.
- No units can 'rout' whilst on a beach.
- If beach engineers suffer casualties they cannot advance next move.
- Beach defenders in prepared positions do not require morale tests.

Beach morale markers in use.

Supporting Naval Gunfire and Air Support

18.21 As soon as a naval gunfire OP team can be landed it may attempt to establish radio contact. Throw three D6. Any throw of 1 means no contact can be made. Otherwise the OP team can direct two shots of 105mm observed IDF per move. An 'air link' OP team takes the same test for contact. It will normally operate without a vehicle if in the assault landing waves.

18.22 From 1944 Allied forces may combine both roles in one OP team as a joint assault signal company (JASCO). A JASCO may only control one type of fire (artillery or aircraft) each move.

Rocket Firing Landing Craft

18.23 Rocket barrages fired from special support landing craft were designed to saturate the beach area and pin down defenders. This fire may be used only once in a game. The procedure is:

1. Using the 12" grid pointed in the direction of fire, place the grid so the centre square (5) is over an enemy beach defence target.
2. Throw a D10 and the square with this number becomes the new point of aim. Reposition the grid.
3. Throw a D6 for every enemy unit, weapon or OP team in open positions under the grid. 4, 5, or 6 = a hit.
4. Test as for multiple mortar fire (12 points table 5) with any deductions.

Rockets have no effect on bunkers and casemates.

18.24 All units under the grid, including those in bunkers and casemates, may not move, fire or observe in their next turn.

Fighter pilot 'Pips' Priller attacks Sword Beach on D-Day, one of a handful of German air attacks. Below: a British escort carrier sails into the wind.

These rules are for players who wish to extend the 'sea' element of amphibious landings or fight small coastal actions. They are not intended for large scale naval actions, which in 20mm, would require a large hall rather than a table top!

Movement

19.1 To allow larger supporting vessels on table, such as destroyers and even escort aircraft carriers, our extension of the table top 'sea' terrain represents a greater distance than its land equivalent. Assuming that the beach or coast lies across the width of the table (normally between 60" and 72") the larger vessels may only be deployed if there is sufficent sea terrain (see **Table 25**).

19.2 A vessel's tightest turning circle has the same diameter as its length.

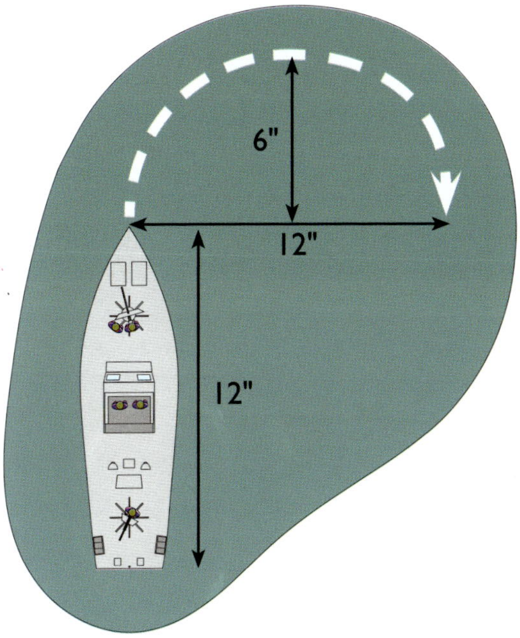

British ML Turning Circle

A British motor launch (ML) glides through calm seas. The ML was a much-used, multi-purpose vessel used for everything from commando raids to escort duties. It was normally armed with a quick-firing 2pdr (40mm) 'pom pom' gun and a 20mm Oerlikon autocannon.

Table 25: Vessel Movement and Crew

Vessel	Maximum Movement Per Turn	Minimum Distance From Beach / Coast*	Maximum Crew***
Small landing craft, barges	Twice own length	May land on beaches	2
Motor launches and patrol boats	Twice own length	May land on beaches**	5
LCT/LCI/LST or similar	Own length	May land on beaches	5 (10 for LST)
Trawlers, fishing boats etc	Own length	6"	5
Motor MG / torpedo boats	Three times own length	6"	5
Gun boat (Insect Class etc)	Own length	18"	12
Destroyer#	Own length	48"	14
Assault troop ship	No movement	60"	12
Escort Carrier	No movement	72"	18

* Example: On a 12' (144") long table, 5' (60") is taken up by land, leaving 7' (84") of sea. Therefore, an escort carrier may be deployed within the last 12" of the 'sea' end of the table.

** Non landing craft need 4, 5 or 6 on a D6 to re-float after beaching.

*** Minimum gun crew (all vessels) = 1 figure. Minimum bridge crew (all vessels) = 1 figure.

\# Destroyers (due to space) will normally remain stopped. If not, they may only move across the width of the table.

Gunfire Ranges and Achieving Hits

19.3 Ranges across sea and other large bodies of water (including those for coastal guns) are simplified into 24" sections that match the popular terrain squares. The firing gun and the target (or its base if an aircraft) must both lie within the maximum range.

Table 26: Naval Gunfire Ranges

Gun	Maximum Range	Minimum Range
150mm (6") and larger	4 sections (96")	1 section (24")
75mm (3") to 149mm (5.9")	3 sections (72")	None
20mm to 74mm	1 section (24")	None

Use Basic Rules ranges and firing procedure for MGs in coastal positions and on vessels firing at land / sea level targets, but deduct 1 from effect throw for MGs on vessels

Table 27: Naval Gunfire Ranges

Target	D6 Throw To Hit
Small landing craft (LCA / LCM etc) and similar sized vessels	6 (then throw a D6 for casualties). Throw a further D6 for each vehicle / weapon. 6 = destroyed.
Motor MG / torpedo boats	6 if moving at two lengths or more (if less: 5 or 6).
Larger landing craft/ other vessels / destroyers	5 or 6 (then throw a D6 for casualties). Throw a further D6 for each vehicle / weapon. 6 = destroyed.
Carriers	4, 5 or 6

19.4 For all vessels hit (other than small landing craft and similar) throw a D6 and check effects set out in **Table 28.**

Bombing

19.5 Aircraft carrying out anti-shipping or carrier-borne bombing attacks are simplified into the following types and armament. They each represent six aircraft.

Dive bombers / carrier bombers: 3 bombs	Light bombers: 6 bombs

19.6 Identify the target to be bombed. The aircraft's base must be placed within 6" of the vessel it is attacking and AA fire range is measured to this position.

19.7 An escort carrier may carry three aircraft. These may be fighters, bombers, or (more commonly) a mixture of both. Only aircraft which historically served on carriers may be used.

A German patrol boat armed with two 20mm autocannon.

The Effects of Guns and Bombs

19.8 For each gun firing, or bomb dropped, throw a D6. Aircraft normally drop to low altitude to attack. Deduct 1 from the throw if an aircraft bombs at medium altitude.

Table 28: Damage from Gunfire and Bombs

D6 Throw	Effect
1	Hit on the vessel's bridge (the command point).
2	Hit on any light AA gun (including AA MG / Oerlikon).
3	Hit on forward turret / gun / LCT hold /LST deck/ landing ship deck / carrier's flight deck (no aircraft may take off until repaired).
4	Hit on rear turret / gun / LCT hold / LST deck / landing ship deck / carrier's flight deck (no aircraft may land until repaired).
5	Hit on medium AA gun (Bofors/pom pom etc) or any other HE weapon (If more than one gun is carried throw D6 to decide which is hit).
6	Hit on engine room. Vessel stops (repair in one D6 moves).
Then throw a D6 for crew / vehicle losses:	
1, 2 or 3	One figure casualty
4, 5 or 6	Two figure casualties
Throw a further D6 for any ships' gun that is hit, or any vehicle, towed gun or aircraft in the target area. 6 = destroyed.	

19.9 If a vessel cannot crew its bridge it will stop and cannot get underway again until the bridge is manned. Obviously, guns with no crew cannot continue firing. If all figures on board a vessel become casualties it sinks.

Carriers and Air Attacks

19.10 Carrier aircraft can spend up to five moves airborne in ground attack or fighter cover (fighters only) role.

19.11 To land or take off takes one full move. To re-bomb and refuel aircraft takes one full move on board the carrier (a minimum of two crew figures are needed to carry this out).

19.12 If the flight deck is damaged by gunfire or bombs it takes a minimum of two crew members to repair in one D6 moves.

Carrier Fighter Cover

19.13 If a carrier fighter is airborne as air cover throw one D6 before an enemy bomber arrives on table. Deduct this number of bombs from the aircraft's bomb load. This simulates the number of aircraft shot down or driven off before arriving over the target.

Naval Anti-Aircraft Fire

19.14 Throw one D10 per anti-aircraft gun and consult **Table 29**.

Table 29: Naval Anti-Aircraft Firing

Unit / Weapon	Target Altitude	To Drive Off
AA MG(s) Oerlikon (20mm)	Low only	1 or 2
Single-barrel 30mm and larger AA guns (Bofors etc)	Low / medium	1 or 2
Multiple 30mm and larger AA gun (pom pom etc)	Low / medium	1, 2, 3 or 4

19.15 Any aircraft driven off may not attack for one full move. If driven off again the aircraft is considered shot down.

Bofors AA gun (below) and Swordfish light bomber (right) onboard an escort carrier. Bottom: The Swordfish in action.

Terrain, Climate and Darkness

Extreme types of terrain and climate played a major role in WWII and night fighting became more common. These rules reflect their impact on tactical warfare.

Hedgerows

20.1 Tall and dense hedgerows, such as the Normandy 'Bocage', offer total concealment for all vehicles observed from ground level and may be classed as hard cover. AFVs crossing high hedgerows are more vulnerable to AT weapons using reserved fire (see **rule 8.42** and **Table 10**).

Desert

20.2 Some areas of the table may be declared as soft sand. Throw a D6 for each vehicle crossing this area per move. 6 = bogged down. Throw again for the number of moves to break free.

Jungle

20.3 Troops cutting a new trail through jungle move at a maximum speed of 3" per move. AFVs may force a trail. Deduct 4" from normal maximum cross-country movement.

20.4 Treat jungle in the same way as woods for all ground and air observation (see **rules 2.11 and 15.24**).

Mountains

20.5 Troops and pack animals climbing very steep terrain may move at a maximum of 4" per move. Cavalry move at a maximum speed of 8".

20.6 Only specialist mountain troops or commandos / rangers may climb and descend sheer rock faces, at a speed of 2" per move. A mountain battalion may improvise a hoist to raise or lower heavy weapons and sections of mountain gun (in three loads) up and down sheer faces at the same speed. The hoist takes two figures two full moves to erect.

20.7 Mountain guns (normally 75mm) may be split into three loads: trail, barrel / wheels and ammunition, carried by three mules or ponies. Pack animals carrying weapons only become casualties if a throw for HE effect destroys the weapon. A mountain gun takes one full move to assemble or dismantle.

20.8 Radio reception can be badly affected by mountainous terrain. All D6 throws for radio contact will fail for scores of 1 or 2.

Swamp

20.9 Swamps and marshy ground are impassable to all vehicles. Infantry and cavalry may cross swamps at ½ cross country speed.

Bocage hedgerows are passable by AFVs but treat them as permanent obstacles for gapping by engineers. This reduces the chances of attacking tanks being hit by reserved anti-tank fire.

Playing Tip

Mud and Floods

20.10 Heavy mud and flooded areas slow all vehicles except tracked amphibians (e.g. Weasels, LVTs). In mud, reduce infantry, cavalry and vehicle movement by a third, except on hard-surface roads (agreed before the game). Cyclists move at infantry rates. Floods are impassable to infantry and all non-amphibious units.

Snow and Ice

20.11 All cross country vehicle movement is reduced by half, including over any rivers or lakes declared as frozen. Vehicle road speed is reduced to cross country rates. Reduce all other movement by a third.

20.12 Ski troops move at 9" cross country. Powered sledges (e.g. Aerosans) move at 15" on level cross country terrain (no movement up hills), frozen rivers and lakes. They may cross roads, but not move along them.

Mist and Fog

20.13 Mist and fog reduce all observation ranges to a maximum of 12". Mist is most likely to occur at the beginning or end of games that represent a full day of fighting.

Night Fighting

20.14 In darkness all elements' movement is reduced by half. Halve all observation ranges when locating moving units or infantry carrying out engineering tasks and reduce to a quarter if units are static. Normal observation ranges apply if targeting any enemy units that fired in the turn before.

- Deduct 1 from all fire effect throws when firing at night.
- Deduct 1 from all morale throws for defending regular and poor units attacked at night. Troops classed as poor may not attack.
- AFVs must test before advancing unless following infantry. Throw a D6. An AFV will advance for a score of 5 or 6, or 4, 5 or 6 if Japanese.

20.15 Light mortars (2", 60mm etc) may fire one parachute flare per move. Identify a target out to a maximum range of 18". Throw a D6. A score of 2, 4, or 6 means the target may be observed and fired at as if in daylight. Burning buildings automatically light up any units within 6" in all directions and these may also be observed and fired at normally.

Weather, Terrain and Aircraft

20.16 Players should agree if weather prevents air support and observation before the game. Aircraft may not be used in mist, fog, cloudy conditions or during snow storms. Aircraft at low altitude in mountains throw a D6 every move. 6 means the aircraft crashes.

Operation Martlet - A Brigade Level Scenario with Advanced Rules

THE BRITISH ATTACK GERMAN-HELD TERRITORY IN NORMANDY, 26TH JUNE 1944

Background

It's 20 days after D-Day. The 49th (Polar Bears) Division has been ordered to attack in support of a forthcoming major British offensive: Operation Epson. The much smaller Operation Martlet aims to capture important high ground to protect the eastern flank of the Epson offensive, but by the third day of fighting the ultimate objective is still in enemy hands. Two infantry battalions and a motor battalion are now available for the attack, supported by tanks of 8th Armoured Brigade. They are faced by units of 12th SS Panzer Division, including tanks and nebelwerfer multiple rocket launchers. The British don't realise that the German forces are about to launch their own attack.

Objectives

To win, the British force must reach the E-W road at Le Manoir and get at least one unit onto the high ground north of Rauray within 15 moves. The German force has to stop the British achieving these objectives and will win if, within 15 moves, they have held the British at the E-W road at La Grand Ferme within the same game period. Any other result will be classed as a draw.

Forces

British Forces		**Orbat Page Number**
70th Infantry Brigade (commanding elements of other brigades)	Brigade HQ	8
1st Tyneside Scottish Regiment + C Squadron 4th / 7th Dragoon Guards	Infantry battalion / Sherman squadron	8 / 6
7th Duke of Wellington's Regiment + C Squadron Nottinghamshire Yeomanry	Infantry battalion / Sherman squadron	
12th Kings Royal Rifle Corps + A Squadron 24th Lancers	Motor battalion / Sherman squadron	7 / 6
147th (SP) Field Regiment	Artillery regiment (DSF)	7
143rd Field Regiment	Artillery regiment (predicted and GSF)	9
Reinforcements (enter at point B on move 9)		
HQ and A and B Squadrons Notts Yeomanry	Regimental HQ and two Sherman squadrons	6

German Forces		**Orbat Page Number**
Kampfgruppe HQ		12
SS Panzer Grenadier Regiment 26 3rd Battalion / 13th, 14th, 15th and 16th Companies	Armoured panzer grenadier battalion and support companies	12 / 13
8th Company SS Panzer Regiment 12	Panzer IV company	14
1st (SP) Battalion SS Artillery Regiment 12	Self-propelled artillery battalion	14
SS Werfer Battalion 12	Nebelwerfer (OP in Kubelwagen)	1 x MRL + tow
Counter-Attack Force (enter at point C on move 4)		
2nd Company SS Panzer Regiment 12	Two Panther tanks	14
6th Company SS Panzer Regiment 12	Panzer IV company	
SS Reconnaissance Battalion 12	Reconnaissance battalion	See back cover

Morale

All British units are regular. The British force test by infantry battalions and tank squadrons. All German infantry are regular. All German tank units are elite. The German force tests by infantry battalions and tank companies.

Deployment, Artillery and Aircraft

British: The Tyneside and KRRC groups enter from move 1 at point A (Tyneside first). The Duke of Wellington group enters from move 1 at point B. All guns are off-table. The British artillery is allowed three moves of predicted fire (both regiments). The 143rd Field Artillery Regiment is then only available for requested GSF fire. One Typhoon fighter bomber may be pre-ordered.
German: the initial force may deploy anywhere south of the dotted red line, with troops and support weapons dug-in and tanks hull down. Units may be concealed by drawing positions on a command map or using markers. The SP artillery battalion OP is allowed three moves of pre-registered fire.

Introducing the Advanced Rules Into a Refight of Operation Martlet

Command and Control (advanced rules 10.1 and 10.2)
Both players draw command maps. The British decide to ring the three farms and the high ground north of Rauray as their main objectives, with arrows showing the three infantry battalions' routes of advance. Both sides have clearly identified highest level HQs that must be on-table throughout the game. The British written orders also include the move on which an air attack will (hopefully) arrive on table. Orders are simple and to the point: e.g.

7th DWR/C Squadron Notts Yeomanry:	*Attack and capture St Nicholas farm, then proceed to capture and hold high ground north of Rauray.*

The Germans aggressively (and ambitiously!) ring the T-junction in Fontenay-le-Pesnil as their main counter-attack objective. As defenders, they mark on concealed units in, or close to, all three farms (no need for markers). Their orders are very clear:

1st Company SS Pz Gren Bn 26 + MMG and 1 Pz IV (St Nicholas Farm):	*Hold the farm at all costs. No surrender!*

Ammunition Supply (advanced rules 11.1)
Both sides agree to restrict tank ammunition to five rounds per vehicle. Therefore both sides will deploy two supply trucks.

Specialist Small Arms (advanced rule 13.1)
Both players agree that the experienced and well-equipped SS panzer grenadiers may benefit from the German LMG rule but will only add 1 to fire effect to represent their losses during the earlier fighting.

Specialist Artillery (advanced rules 14.1, 14.2 and 14.8)
British: one regiment of SP artillery (147th) is available for direct support fire (DSF) throughout the attack, but 143rd Field Regiment is only available for three moves of predicted fire and for GSF during the rest of the game. Three OP tanks are deployed, one with each infantry battalion.
The British Fire Plan:

Moves 1-3 predicted fire:	*Two moves of fire against the three farms (as geographical targets with appropriate deductions for effect). 4 x 25pdrs against each target. One move of smoke against the same targets (same number of guns).*

German: the German force is allowed three points as pre-registered targets for the 1st (SP) SS Artillery Battalion only. The German player selects the pre-registered points and marks them on the command map. Two OPs may be deployed in total, one in a half-track controlling the guns and one, in a Kubelwagen, controlling the Nebelwerfer battery.

8th Company Pz IVs face Notts Yeomanry Shermans near St Nicholas farm.

Aircraft (advanced rule 15.4)
The Britsh pre-order an air attack by a rocket-armed Typhoon fighter-bomber to arrive on move 8. Orders and targets must be written down on move 4. The British player will then have to test on move 8 to see if the Typhoon arrives on time.

Move 2. The defenders of St Nicholas farm are pounded by British artillery. Deductions for hard cover and predicted fire against a geographical feature mean casualties are relatively light.

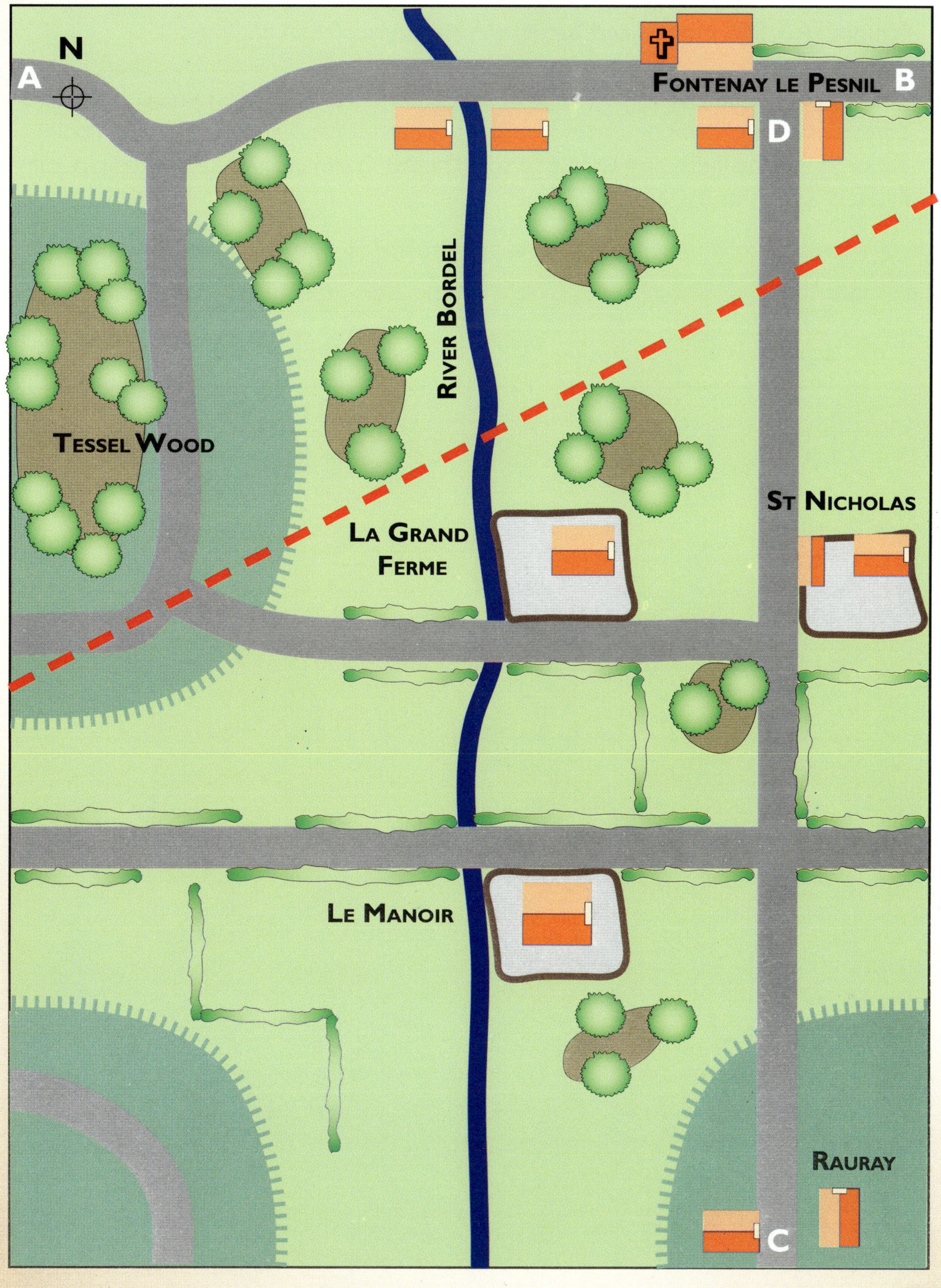

The two buildings in Fontenay-le-Pesnil at D are ruins, all other buildings are intact. All hedges are tall, thick bocage. The River Bordel is a linear obstacle. All woods are open.

Rapid Fire! is supported by a website:

http://www.rapid-fire.uk.com

On the site you will find all the latest information on ***Rapid Fire!*** including vehicle and gun charts, playing aids, scenario and modelling articles and even more photographs!

The models pictured in this book are from the authors' own collections and many have seen extensive action over the years. Manufacturers of the figures and vehicles include: Raventhorpe Miniatures and 'Ready to Roll'; Airfix, Hasegawa, Italeria, Esci and Revell (plastic models:), and Skytrex, Britannia, Combat Ready, AB Figures, FAA, S&S Models and Battlefield Miniatures (resin and metal).

Buildings are mainly Sentry Models or scratch-built, as is the terrain, with construction and painting carried out by the authors and Sue Rumford, Malcolm Taylor, Tony Chadburn and Dale Rawson. The AB Germans were painted for Colin by Steve Goodhand.

Operation Martlet. SS Recce Battalion 12 joins the battle for St Nicholas farm on move 9.